THE POTENTIAL IMPACT OF
HIGH-END CAPABILITY COMPUTING
ON FOUR ILLUSTRATIVE FIELDS OF
SCIENCE AND ENGINEERING

Committee on the Potential Impact of High-End Computing on
Illustrative Fields of Science and Engineering

Division on Engineering and Physical Sciences

Division on Earth and Life Sciences

NATIONAL RESEARCH COUNCIL
OF THE NATIONAL ACADEMIES

THE NATIONAL ACADEMIES PRESS
Washington, D.C.
www.nap.edu

THE NATIONAL ACADEMIES PRESS   500 Fifth Street, N.W.   Washington, DC 20001

NOTICE: The project that is the subject of this report was approved by the Governing Board of the National Research Council, whose members are drawn from the councils of the National Academy of Sciences, the National Academy of Engineering, and the Institute of Medicine. The members of the committee responsible for the report were chosen for their special competences and with regard for appropriate balance.

This study was supported by Contract No. NCO-0610176 between the National Academy of Sciences and the National Science Foundation. Any opinions, findings, conclusions, or recommendations expressed in this publication are those of the author(s) and do not necessarily reflect the views of the organizations or agencies that provided support for the project.

Library of Congress Cataloging-in-Publication Data

National Research Council (U.S.). Committee on the Potential Impact of High-End Computing on Illustrative Fields of Science and Engineering
  The potential impact of high-end capability computing on four illustrative fields of science and engineering / Committee on the Potential Impact of High-End Computing on Illustrative Fields of Science and Engineering, Division on Engineering and Physical Sciences, Division on Earth and Life Sciences.
     p. cm.
  Includes bibliographical references.
  ISBN 978-0-309-12485-0 (pbk.) — ISBN 978-0-309-12486-7 (pdf) 1. High performance computing—United States. 2. Supercomputers—United States. 3. Science—Data processing. 4. Engineering—Data processing. I. National Research Council (U.S.). Division on Engineering and Physical Sciences. II. National Research Council (U.S.). Division on Earth and Life Studies. III. Title.
  QA76.88.N35 2008
  502.85—dc22
                                     2008037012

Additional copies of this report are available from the National Academies Press, 500 Fifth Street, N.W., Lockbox 285, Washington, DC 20055; (800) 624-6242 or (202) 334-3313 (in the Washington metropolitan area); Internet, http://www.nap.edu.

Copyright 2008 by the National Academy of Sciences. All rights reserved.

Printed in the United States of America

THE NATIONAL ACADEMIES
Advisers to the Nation on Science, Engineering, and Medicine

The **National Academy of Sciences** is a private, nonprofit, self-perpetuating society of distinguished scholars engaged in scientific and engineering research, dedicated to the furtherance of science and technology and to their use for the general welfare. Upon the authority of the charter granted to it by the Congress in 1863, the Academy has a mandate that requires it to advise the federal government on scientific and technical matters. Dr. Ralph J. Cicerone is president of the National Academy of Sciences.

The **National Academy of Engineering** was established in 1964, under the charter of the National Academy of Sciences, as a parallel organization of outstanding engineers. It is autonomous in its administration and in the selection of its members, sharing with the National Academy of Sciences the responsibility for advising the federal government. The National Academy of Engineering also sponsors engineering programs aimed at meeting national needs, encourages education and research, and recognizes the superior achievements of engineers. Dr. Charles M. Vest is president of the National Academy of Engineering.

The **Institute of Medicine** was established in 1970 by the National Academy of Sciences to secure the services of eminent members of appropriate professions in the examination of policy matters pertaining to the health of the public. The Institute acts under the responsibility given to the National Academy of Sciences by its congressional charter to be an adviser to the federal government and, upon its own initiative, to identify issues of medical care, research, and education. Dr. Harvey V. Fineberg is president of the Institute of Medicine.

The **National Research Council** was organized by the National Academy of Sciences in 1916 to associate the broad community of science and technology with the Academy's purposes of furthering knowledge and advising the federal government. Functioning in accordance with general policies determined by the Academy, the Council has become the principal operating agency of both the National Academy of Sciences and the National Academy of Engineering in providing services to the government, the public, and the scientific and engineering communities. The Council is administered jointly by both Academies and the Institute of Medicine. Dr. Ralph J. Cicerone and Dr. Charles M. Vest are chair and vice chair, respectively, of the National Research Council.

www.national-academies.org

COMMITTEE ON THE POTENTIAL IMPACT OF HIGH-END COMPUTING ON ILLUSTRATIVE FIELDS OF SCIENCE AND ENGINEERING

JOHN W. LYONS, National Defense University, *Chair*
DAVID ARNETT, University of Arizona
ALOK N. CHOUDHARY, Northwestern University
PHILLIP COLELLA, Lawrence Berkeley National Laboratory
JOEL L. CRACRAFT, American Museum of Natural History
JOHN A. DUTTON, Storm Exchange, Inc., and Pennsylvania State University (retired)
SCOTT V. EDWARDS, Harvard University
DAVID J. ERICKSON III, Oak Ridge National Laboratory
TERESA L. HEAD-GORDON, University of California at Berkeley
LARS E. HERNQUIST, Harvard-Smithsonian Center for Astrophysics
GEORGE E. KELLER II, MATRIC
NIPAM H. PATEL, University of California at Berkeley
MARY E. REZAC, Kansas State University
RONALD B. SMITH, Yale University
JAMES M. STONE, Princeton University
JOHN C. WOOLEY, University of California at San Diego

Staff

SCOTT WEIDMAN, Study Director, Division on Engineering and Physical Sciences
BARBARA WRIGHT, Administrative Assistant

Preface

The study that led to this report was called for in the President's Fiscal Year 2006 Budget. In accordance with that call, the study was commissioned by the National Coordination Office (NCO) for Networking and Information Technology Research and Development (NITRD), which supports the NITRD Subcommittee, which operates under the White House National Science and Technology Council to coordinate federal investments in networking and information technology research and development. NCO efforts were guided by a steering group appointed by the NITRD Subcommittee and input from the White House Office of Science and Technology Policy and the White House Office of Management and Budget. The study was conducted by a committee (see Appendix A) constituted under the National Research Council's (NRC's) Division on Engineering and Physical Sciences and its Division on Earth and Life Sciences.

This report addresses the following charge:

The study will develop a better understanding of the potential scientific and technological impact of high-end capability computing in four illustrative fields of S&E of interest to the federal government. More specifically, the study will

(a) Review the important scientific questions and technological problems identified for those fields in other sources (e.g., decadal surveys);
(b) Identify the subset of those important questions and problems for which an extraordinary advancement in understanding is difficult or impossible without high-end capability computing;
(c) Identify some of the likely impacts of making progress on as many of the scientific questions and technological problems identified in (b) as possible and the contribution that high-end capability computing can make to this progress;
(d) Discuss some of the most significant ramifications of postponing this use of high-end capability computing in order to capitalize on the decreasing cost of computing over time;
(e) Identify the numerical and algorithmic characteristics of the high-end capability computing requirements needed to address the scientific questions and technological problems identified in (b); and

(f) Categorize the numerical and algorithmic characteristics, specifically noting those categories that cut across disciplines. This task shall be done in a way that can later be used to inform design and procurements of high-end capability systems.

This list of tasks is not in priority order. Tasks (a), (b), (e), and (f) are considered to be the most important and essential for the study's success.

When the NCO asked the NRC to undertake this study, it requested that the committee consider, for specificity, the potential impacts of high-end capability computing (HECC) on four illustrative fields chosen from the following broad areas of science and engineering that are of importance to the federal government: physics and/or astronomy, the geosciences, chemistry and/or chemical engineering, and the biological sciences and/or biomedical engineering.

NRC staff from the Division on Engineering and Physical Sciences and the Division on Earth and Life Sciences developed the following criteria for selecting illustrative fields from within those broad areas:

- Each field selected for study should have relatively well-defined goals, and the goals should be documented in consensus reports so that the current study need not identify the major research challenges from scratch.
- The fields selected should be broad enough to be of widespread interest yet limited enough to be fairly represented through the study's process, which is outlined below.
- Collectively, the four fields selected should span the most important areas of interest to the NITRD community: the geosciences, defense, health, space, energy, commerce, and basic research.
- The four fields selected should illustrate a range of comfort, usage, and acceptance of HECC approaches to investigation.

By meeting these four criteria, it was felt that the fields selected would adequately illustrate the range of science and engineering research topics that the NITRD program encompasses and most likely raise a broad set of HECC-related issues that the NITRD program addresses. The committee did not believe it was necessary, or even desirable, to exhaustively survey the entire community supported by NITRD's HECC programs, nor was that tactic called for in the charge.

Based on these criteria, NRC staff held a number of discussions with experts in physics, astronomy, the geosciences, chemistry, chemical engineering, and biology and with representatives of the NITRD community. An attempt was made to include at least one field that is not obviously dependent on HECC. As a result, it was decided to focus the study on the following four illustrative fields of science and engineering:

- Astrophysics,
- The atmospheric sciences,
- Evolutionary biology, and
- Chemical separations.

While this study does identify the potential impact of HECC in these four fields, and thus implicitly points out some funding opportunities, that is not its real goal, and this study is no substitute for competitive review of specific proposals. Rather, the study is meant to define the sort of examination that any field or federal agency could undertake in order to analyze the HECC infrastructure it needs to support progress against its research goals, within the context of other means of attacking those goals.

In the course of such an examination, a field or agency will methodically evaluate its ability to take advantage of HECC infrastructure and in so doing will identify the components of the infrastructure that are needed to enable progress. By defining the self-examination, the study also implies a vision for the environment that will best profit from HECC investments, and it provides guidance for federal policy makers who must weigh competing requests for those funds.

The committee met four times in the course of its study; agendas are included in Appendix B. Of special note was the second meeting, which included four small workshops in parallel (one for each of the four fields) to obtain community input on Tasks (a)-(e) of the study's charge. The list of participants in those workshops is also in Appendix B.

The goal of these information-gathering workshops was not to canvass broadly—the study was not meant to identify de novo the leading questions in the four fields and relied on other sources—but to have a focused discussion with a small number of leaders in each field plus federal scientists and engineers. The workshops discussed whether a draft list of major research challenges adapted from other sources would be adequate for the purposes of this study; identified which of those challenges are critically dependent on HECC, and in what way(s); evaluated the abilities of the four fields to capitalize on HECC for addressing those challenges; and identified critical gaps in each field's capabilities for realizing the opportunities presented by HECC. The results of these discussions are captured in Chapters 2-5.

The committee's third and fourth meetings focused on crosscutting elements of the charge, primarily Task (f). Those discussions led to Chapter 6, which presents and analyzes the HECC capabilities and needs for the four fields, and Chapter 7, which essentially provides a decision process for fields, agencies, and federal policy makers to use in assessing the potential value of HECC investments.

Acknowledgments

This report has been reviewed in draft form by individuals chosen for their diverse perspectives and technical expertise, in accordance with procedures approved by the National Research Council's Report Review Committee. The purpose of this independent review is to provide candid and critical comments that will assist the institution in making its published report as sound as possible and to ensure that the report meets institutional standards for objectivity, evidence, and responsiveness to the study charge. The review comments and draft manuscript remain confidential to protect the integrity of the deliberative process. We wish to thank the following individuals for their review of this report:

THOM H. DUNNING, University of Illinois at Urbana-Champaign,
JOSEPH FELSENSTEIN, University of Washington,
JAMES HACK, Oak Ridge National Laboratory,
DAVID M. HILLIS, University of Texas-Austin,
JOHN P. HUCHRA, Harvard-Smithsonian Center for Astrophysics,
ROBERT F. LUCAS, University of Southern California,
CHRISTOPHER McKEE, University of California at Berkeley,
DANIEL A. REED, Microsoft Corporation,
RICHARD B. ROOD, University of Michigan, and
JEFFREY SIIROLA, Eastman Chemical Company.

Although the reviewers listed above have provided many constructive comments and suggestions, they were not asked to endorse the conclusions or recommendations nor did they see the final draft of the report before its release. The review of this report was overseen by Margaret H. Wright of New York University. Appointed by the National Research Council, she was responsible for making certain that an independent examination of this report was carried out in accordance with institutional procedures and that all review comments were carefully considered. Responsibility for the final content of this report rests entirely with the authoring committee and the institution.

The committee also acknowledges the valuable contribution of the following individuals, who provided input at the meetings on which this report is based:

TOM ABEL, Stanford University,
ARDEN BEMENT, National Science Foundation (NSF),
JOAN BRENNECKE, University of Notre Dame,
ANTONIO J. BUSALACCHI, University of Maryland,
ANNE CHAKA, National Institute of Standards and Technology,
DANIEL DRELL, Department of Energy,
SEAN EDDY, Howard Hughes Medical Institute,
SERGEY GAVRILETS, University of Tennessee, Knoxville,
BRIAN GROSS, NOAA Geophysical Fluid Dynamics Laboratory,
JAMES HACK, Oak Ridge National Laboratory,
SALLY HOWE, National Coordination Office for Networking and Information Technology Research and Development
JOEL KINGSOLVER, University of North Carolina,
JOHN MARBURGER, Office of Science and Technology Policy,
CHRISTOPHER McKEE, University of California at Berkeley,
EVE OSTRIKER, University of Maryland,
JOEL PARRIOTT, Office of Management and Budget,
DANIEL ROKHSAR, Lawrence Berkeley National Laboratory,
ED SEIDEL, Louisiana State University,
NIGEL SHARP, NSF,
JEFFREY SIIROLA, Eastman Chemical Company,
RICK STEVENS, Argonne National Laboratory,
GEORGE STRAWN, NSF,
ALEX SZALAY, Johns Hopkins University,
SIMON SZYKMAN, National Coordination Office for Networking and Information Technology Research and Development
CHAOWEI YANG, NASA Applied Sciences Program, and
MANFRED ZORN, NSF.

The committee is also grateful for the contributions of James McGee and Ann Reid of the National Academies staff, who assisted in the early stages of this study.

Contents

SUMMARY   1

1   PROBLEM DEFINITION AND HISTORY   9
     Introduction, 9
     History of High-End Computing, 11
     Current State of High-End Capability Computing, 13

2   THE POTENTIAL IMPACT OF HECC IN ASTROPHYSICS   15
     Introduction, 15
     Major Challenges in Astrophysics, 17
     Major Challenges That Require HECC, 18
     Methods and Algorithms in Astrophysics, 30
     HECC for Data Analysis, 32
     Realizing the Potential Impact of HECC on Astrophysics, 33
     References, 35

3   THE POTENTIAL IMPACT OF HECC IN THE ATMOSPHERIC SCIENCES   37
     Introduction, 37
     Major Challenges in the Atmospheric Sciences, 39
     Computational Challenges in the Atmospheric Sciences, 46
     The Need for HECC Resources to Advance the Atmospheric Sciences, 57
     Conclusion: Earth in a Computer, 60
     References, 60

4   THE POTENTIAL IMPACT OF HECC IN EVOLUTIONARY BIOLOGY   63
     Introduction, 63
     Major Challenges of Evolutionary Biology, 64

Major Challenges in Evolutionary Biology That Require HECC, 79
References, 85

5 THE POTENTIAL IMPACT OF HECC IN CHEMICAL SEPARATIONS 89
Introduction, 89
Major Challenges Facing Chemical Separations, 92
Potential Impacts of HECC for Chemical Separations, 96
Current Frontiers of HECC for Chemical Separations, 99
Other Issues That Limit the Value of HECC to Chemical Separations, 103
References, 103

6 NUMERICAL AND ALGORITHMIC CHARACTERISTICS OF HECC 105
THAT WILL BE REQUIRED BY THE SELECTED FIELDS
Numerical and Algorithmic Characteristics of HECC for Astrophysics, 105
Numerical and Algorithmic Characteristics of HECC for the Atmospheric Sciences, 108
Numerical and Algorithmic Characteristics of HECC for Evolutionary Biology, 110
Numerical and Algorithmic Characteristics of HECC for Chemical Separations, 112
Categorization of Numerical and Algorithmic Characteristics of HECC Needed in the
    Four Selected Fields, 114
Crosscutting Challenges from Massive Amounts of Data, 116
Crosscutting Challenges Related to Education and Training, 118

7 CONCLUSIONS 121
Supporting High-End Computational Research, 121
The Need for Continuing Investment in HECC, 123
Classes of Numerical and Algorithmic Challenges, 125
Human Resources, 125
Lessons Learned for Fields That Might Perform Similar Studies, 126

APPENDIXES

A Biographical Sketches of Committee Members 129
B Agendas of Committee Meetings 133
C Glossary 141

THE POTENTIAL IMPACT OF HIGH-END CAPABILITY COMPUTING ON FOUR ILLUSTRATIVE FIELDS OF SCIENCE AND ENGINEERING

Summary

INTRODUCTION

Many federal funding requests for more advanced computer resources assume implicitly that greater computing power creates opportunities for advancement in science and engineering. This has often been a good assumption. Given stringent pressures on the federal budget, the White House Office of Management and Budget (OMB) and its Office of Science and Technology Policy (OSTP) are seeking an improved approach to the formulation and review of requests from the agencies for new computing funds. The study that produced this report was commissioned by the Networking and Information Technology Research and Development (NITRD) program, which operates under the OSTP to coordinate federal investments in networking and information technology. The study addressed the charge shown in the Preface.

The study considered, as examples, four fields of science and engineering to determine which of their major challenges are critically dependent on high-end capability computing (HECC). The fields chosen for the study were the atmospheric sciences, astrophysics, chemical separations, and evolutionary biology. The committee found continuing demands from the four fields for more, and more powerful, high-end computing. All four areas rely on HECC to carry out simulations of systems that are too complex to analyze through observation, experiment, or theory. Three of the four areas (the exception being chemical separations) are dealing with very large amounts of data and need HECC to handle them.

"High-end capability computing" means advanced computing that pushes the bounds of what is computationally feasible. While that is often interpreted in terms of raw processing power, users expect to achieve new scientific understanding or engineering capabilities through computation. Processing power is just one means to that end. Computational science and engineering is a systems process, bringing together hardware, software, investigators, data, and other components of infrastructure in order to gain insight into some question. Thus, from the user's perspective, high-end capability computing means whatever sort of advanced, nonroutine computing system is needed to push the computational science or engineering capabilities of a given field. It will always entail more risk and require more innovation than commodity computing, but not necessarily a novel computing platform. This report uses the term "high-end capability computing," or HECC, as shorthand for this nonroutine frontier of computation, the precise definition of which will vary by field.

While this study does identify the potential impact of HECC in these four fields, and thus implicitly identifies some potential funding opportunities, that is not the goal, and this study is no substitute for competitive review of specific proposals. Rather, the study is meant to illustrate the sort of examination that any field or federal agency could undertake in order to analyze the HECC infrastructure it needs to support progress toward its research goals, within the context of other means of pursuing those goals.

SUMMARY OF THE MAJOR CHALLENGES IN THE FOUR FIELDS

Astrophysics

A small sample of some of the most important discoveries in astrophysics made in the past decade includes dark matter and dark energy, exosolar planets, and good evidence for the existence of black holes. These discoveries have positioned the field to address the following major challenges:

1. What is dark matter?
2. What is the nature of dark energy?
3. How did galaxies, quasars, and supermassive black holes form from the initial conditions in the early Universe observed by the Wilkinson Microwave Anisotropy Probe (WMAP) and the Cosmic Backround Explorer (COBE), and how have they evolved since then?
4. How do stars and planets form, and how do they evolve?
5. What is the mechanism for supernovae and gamma-ray bursts, the most energetic events in the known Universe?
6. Can we predict what the Universe will look like when observed in gravitational waves?

To answer the questions posed by Challenges 3-6, advances in HECC are necessary. Challenges 1-2 are limited in the near term by the need for advanced astronomical observations, but these observations will produce so much data that HECC will in any case be needed for their analysis. The current situation for these challenges is described in Chapter 2.

While astrophysics is a computationally mature discipline—that is, it has a long history in the use of computing to solve problems—it would certainly benefit from access to more, and more powerful, HECC resources. The primary computational challenge is associated with the enormous dynamic range in length scales and timescales needed to resolve astrophysical processes. For example, grids as large as 2048^3 are currently used for calculations involving hydrodynamics, but even then the spatial features they are able to resolve are only about a hundredth the size of the computational domain. The availability of systems of 10^5 or more processors will enable much larger calculations (encompassing finer resolution, models of more physical processes, or both) while also making it feasible to perform more complex calculations that couple different models. The community needs support for porting its codes to multicore and petascale environments.

The committee identified some likely ramifications of inadequate or delayed support of HECC for astrophysics:

- The rate of new discovery would be limited.
- Inadequate support for HECC would lead to a failure to optimize investment in expensive experimental and observational facilities.

- Data are likely to be underexploited. Without enough HECC, the data collected by large-scale surveys cannot be properly managed and analyzed, so their full potential cannot be realized.

The Atmospheric Sciences

Weather forecasting and climate simulation require detailed simulations of the atmosphere. The skill and reliability of forecasts has increased markedly since the advent of weather radar, earth-observing satellites, and powerful computers. To look beyond a few hours, we combine powerful computers with the relevant laws of physics converted into mathematical models to predict how the observed present state of the global atmosphere will evolve in the hours and days ahead. We expect that major improvement will be achieved with much higher resolution and sophistication in the numerical models that portray events in the atmosphere and ocean and on the land surface. Numerical weather forecasting, in particular, will soon be able to take account of local features (e.g., lakes, ridges) and local variations in atmospheric moisture content, and the successful modeling of atmospheric variables on these scales should lead to a leap forward in forecast quality. Climate simulation and prediction requires detailed treatments of the physics, chemistry, and biology of the atmosphere, ocean, and land surface. Feedbacks in the integrated Earth system require high spatial resolution, as is the case for weather forecasting, but also additional mechanistic models of chemical and biological interactions.

The major challenges facing the atmospheric sciences are these:

1. Extend the range, accuracy, and utility of weather prediction. [1]
2. Improve our understanding and the timely prediction of severe weather, pollution, and climate events. [1]
3. Improve understanding and prediction of seasonal, decadal, and century-scale climate variation on global, regional, and local scales. [1]
4. Understand the physics and dynamics of clouds, aerosols, and precipitation. [2]
5. Understand the atmospheric forcing and feedbacks associated with moisture and chemical exchange at Earth's surface. [1]
6. Develop a theoretical understanding of nonlinear bifurcation and tipping points in weather and climate systems. [2]
7. Create the ability to accurately predict global climate and carbon-cycle response to forcing scenarios over the next 100 years. [1]
8. Model and understand the physics of the ice ages, including embedded abrupt climate change events such as the Younger Dryas, Heinrich, and Dansgaard-Oeschger events. [2]
9. Model and understand the key climate events in the early history of Earth and other planets. [3]

The committee believes that progress on the challenges marked [1] in this list (Challenges 1-3, 5, and 7) would immediately accelerate with advances in computing capability. Those marked [2] are using or will shortly use current capability computing. For the one marked [3], HECC will probably not play a big role within the next 5 years.

HECC in the atmospheric sciences mainly involves simulations based on the coupled multidimensional partial differential equations of fluid dynamics and heat and mass transfer. The fundamental atmospheric processes are driven by a variety of forces arising from radiation, moisture processes, chemical reactions, and interactions with land and sea surfaces. We need to increase the horizontal mesh resolution by a factor of between 4 and 10 to meet the rising demands from users of weather predictions. That, in turn, necessitates a hundred- to thousandfold increase in computing capability.

Climate simulations are improved largely by the addition of physical processes and the coupling of processes, both of which exacerbate the computational challenge. Compounding that difficulty is the need to compress longer simulated periods into no more than a few months of run time. Among climate scientists, a minimum of 10 simulated years of model integration is typically thought to be needed for credible work. Such a standard, on top of increasing complexity and a need for higher resolution, creates yet more demand for HECC.

Evolutionary Biology

The committee identified the following as the major challenges facing evolutionary biology today:

1. What has been the history of life?
2. How do species originate?
3. How has life diversified across space and time?
4. What determines the origin and evolution of the phenotype?
5. What are the evolutionary dynamics of the phenotype-environment interface?
6. What are the patterns and mechanisms of genome evolution?
7. What are the evolutionary dynamics of coevolving systems?

Observation and experiment continue to be productive modes of inquiry for these major challenges, and because computational evolutionary biology is still young, progress through computational research is still possible with modest computing capabilities. Today, researchers can still investigate many aspects of these major challenges—pose questions, explore relationships and models, and develop algorithms—without reaching the level of complexity that calls for HECC. But for many, that is rapidly changing. Resolving relationships among species, individuals, or genes or analyzing the huge amount of available genomic data are already driving many of these computational efforts toward algorithmic and computational complexity and, thus, the need for HECC resources. Some research into Major Challenges 1, 3, 6, and 7 is already making use of capability computing. Eventually, HECC will be necessary to make progress on all of these challenges, as evolutionary biology relies more heavily on data mining and modeling.

Evolutionary biology is somewhat of a special case because the rapid development of large-scale genomic analysis has enabled radical new approaches, and the field is very much in transition from one based on observation to one based on massive amounts of genomic data. The eventual impacts of HECC are clear and enormous, but the field is only beginning to exploit HECC.

Chemical Separations

The issues facing chemical separations are very different from those discussed in connection with the other fields because it is a field that is dominated by the industrial sector and uses well-understood twentieth-century technologies that would be expensive to reformulate. However, competition in the industry is exerting pressure on U.S. manufacturers to develop technologies for more energy-efficient and environmentally friendly separations processes to better manufacture pharmaceuticals, reduce greenhouse gases in emissions, and increase drinking water supplies that may become scarce in the future. It is very difficult to develop these demanding processes through experimentation alone, and so the time is coming when advanced computation will be more critical for technological advance.

For instance, distillation is highly effective for separating compounds based on differences in their relative volatilities. Yet, because it requires that the mixture be repeatedly vaporized and condensed, it consumes very large amounts of energy. Nonetheless, distillation is by far the most common separation process, used in as much as 80 percent of the most common chemical separations processes, so that optimization of phase equilibria will remain important for the chemical separations industry. Mass-separating agents (MSAs)—solvents, absorbents, adsorbents, membranes, and so on—are also now being used to amplify the separating capability and provide more economical and environmentally friendly solutions. However, the design of new MSA-based systems is severely hindered by the lack of physical property data. HECC has the potential to lead to significant breakthroughs in the development of new MSAs and, thus, markedly reduce the energy consumed by separation systems.

The following are the major challenges in chemical separations found by the committee:

1. How can we predict physical properties at the level of accuracy required for defining the optimal conditions for separating mixtures?
2. How can we design, construct, and produce MSAs with appropriately engineered three-dimensional structures (when needed) that facilitate the rapid and efficient accomplishment of difficult separations?
3. How can we design overall separation systems that incorporate several individual separation units for economically optimal separations of complex mixtures?

There are numerous examples of computational chemistry leading to new understanding of the behavior of chemicals in separation systems, and great potential for further benefits in addressing Major Challenges 1 and 2. Because simulations of molecules of industrial importance and of realistic systems are computationally demanding, it is likely that HECC resources will be required for those applications. Major Challenge 3 calls for the ability to optimize the interplay of multiple separation processes to achieve high-performance separation systems for complex mixtures. This is a very demanding computational task, but much work must be carried out before it can be addressed through HECC.

In summary, HECC can play a transformative role in chemical separations by directing experimentation in more productive directions. It can play that role by providing (1) more accurate phase equilibria data for a wider range of chemical compounds and multicomponent systems—and with greater safety when the separations deal with chemical species that may be toxic or dangerously reactive—and (2) fast screening of, and design information for, candidate MSAs. Ultimately, HECC holds the promise of enabling more optimal design of complete chemical separation processes.

CROSSCUTTING OBSERVATIONS

Chapter 6 gives an indication of the requirements in mathematics, computer science, and computing infrastructure associated with the technical challenges identified in Chapters 2 through 5. For astrophysics, only a small fraction of the algorithms of importance are able to scale well to 10^3-10^4 processors, to say nothing of scaling to even larger platforms. Algorithms, models, and software are therefore needed to enhance scalability for a large number of applications. New models are needed to represent multiscale physics in a way that maps well onto as-yet-undefined new computer architectures. Algorithms for discretization, solution of stiff systems of differential equations, and data management are also critical.

In the long term, the atmospheric sciences also will need new algorithms for discretization, solution of stiff systems of differential equations, and data management, and they will need new models to capture multiscale and multiphysics phenomena. However, for the near term, the field could readily exploit a

10-fold increase in computing capability to improve prediction of severe weather and climate prediction; to better support critical industries such as transportation, energy, and agriculture; and to increase the atmospheric grid resolution of a coupled atmosphere-ocean-sea-ice-biogeochemistry climate model.

Evolutionary biology is not currently limited by HECC resources, but that situation is changing rapidly. Adding more species and making more use of genomic data will quickly drive computational evolutionary biology into the realm of high-end computing. Indeed, scalability problems with many algorithms and the massive amounts of genomic data to be exploited will soon limit evolutionary biology if it does not get adequate HECC resources.

Progress in chemical separations is not currently limited by HECC, but important opportunities would open up from increasing computing capabilities. In the short term, the major HECC requirements are those that would enable simulations to be performed more readily so that their ability to guide experimentation could be exploited more routinely. At present, it is often just easier to perform experiments, because the simulations cannot offer enough predictive power. In the longer term, researchers would like to improve the accuracy of the underlying molecular models for a wider range of materials. An equally desirable capacity would be to enable the convergence of simulated observables so as to attain independent prediction of materials properties over the ranges of pressure and temperature needed for their phase diagrams. Achieving this capability will require algorithms and software that routinely use 10^5-10^6 processors per run, which will in turn require new ideas in mathematical models, numerical algorithms, and software infrastructure.

A common challenge facing three of the committee's fields (excepting chemical separations) is managing and exploiting massive (and increasing) amounts of data. Failure to address this issue—which increasingly requires capability computing—could limit our nation's ability to profit from past and ongoing investments in observation and experiment.

Another common challenge for all four fields is preparing the next generation of researchers, who will push the frontiers of computational science and engineering. Investment in education and training for computational science is needed in all core science disciplines where HECC currently plays, or will play, a larger role in meeting the major challenges. HECC investments should address needs for education and training infrastructure. Students need stronger foundations in mathematics and statistics. Two options for advancing the computational capabilities of our future workforce are (1) imparting stronger computational science skills through the normal curricula of HECC-dependent disciplines and (2) developing a distinct undergraduate or graduate track that produces computational "technologists," professionals whose contribution to the research enterprise is enabling efficient and effective computational approaches rather than personally conducting the research. The first option would impose an additional burden on graduate students, forcing them to make curriculum trade-offs or take more time to complete a degree. The second option requires the definition of career optimizing tracks for computational generalists so they can move smoothly into and out of science domains as the need arises for their expertise, as well as cultural adjustments so that their contributions are recognized and career paths exist.

CONCLUSIONS

The committee members, in spite of their diverse backgrounds and varying degrees of reliance on high-end computing, readily agreed that HECC requires the integration of synergistic elements and should be managed as a system.

Conclusion 1. *High-end capability (HECC) computing is advanced computing that pushes the bounds of what is computationally feasible. Because it requires a system of interdependent com-*

ponents and because the mix of critical-path elements varies from field to field, HECC should not be defined simply by the type of computing platform being used. It is nonroutine in the sense that it requires innovation and poses technology risks in addition to the risks normally associated with any research endeavor.

High-end computational capabilities include whatever mix of hardware, models, algorithms, software, intellectual capacity, and computational infrastructure must be deployed to enable the desired computations. High-end computing platforms are certainly part of that mix, and the most ambitious and progressive computational science may, in many cases, require a new generation of hardware.

In Chapter 7 the committee lists 10 prerequisites if a field is to profit from HECC. These will be useful in evaluating investment opportunities.

Conclusion 2. *Advanced computational science and engineering is a complex enterprise that requires models, algorithms, software, hardware, facilities, education and training, and a community of researchers attuned to its special needs. Computational capabilities in different fields of science and engineering are limited in different ways, and each field will require a different set of investments before it can use HECC to overcome the field's major challenges.*

At the very least, HECC infrastructure will consist of hardware, operating software, and applications software. In addition, there will continue to be a need for data management tools, graphical interface tools, data analysis tools, and algorithms research and development. Disciplines will take advantage of the increased availability of HECC in proportion to how much of the necessary infrastructure has already been created—that is, whether the field has achieved a state of readiness.

Conclusion 3. *Decisions about when, and how, to invest in HECC should be driven by the potential for those investments to enable or accelerate progress on the major challenges in one or more fields of science and engineering.*

Once a decision is made to invest in computational resources, that investment must provide all the elements of infrastructure that are needed by the fields likely to use the resource.

Conclusion 4. *Because the major challenges of any field of science or engineering are by definition critical to the progress of the field, underinvestment in any of them will hold back the field.*

Optimum progress across all the major challenges will be achieved if all modalities of research—theoretical, experimental, and computational—are supported in a balanced way.

In many cases, HECC capabilities must continue to be advanced to maximize the value of data already collected or investments already made in experimental facilities. For instance, remote sensing projects under way in astrophysics and the atmospheric sciences will produce quantities of data that cannot be utilized by those fields without commensurate progress in analytical capabilities.

Conclusion 5. *The emergence of new hardware architectures precludes the option of just waiting for faster machines and then porting existing codes to them. The algorithms and software in those codes must be reworked.*

There do not yet exist productive and easy-to-use programming methodologies or low-level blocks of code that can take full advantage of multicore processors. Multicore parallelism is unfamiliar to many commercial software developers, and it also requires different sorts of parallel algorithm development.

Conclusion 6. *All four fields will need new, well-posed mathematical models to enable HECC approaches to their major challenges. Astrophysics and the atmospheric sciences share two needs: one for new ways to handle stiff differential equations and one for continuing advances in multiresolution and adaptive discretization methods. Astrophysics and chemical separations also share two needs: one for accurate and efficient methods for evaluating long-range potentials that scale to large numbers of particles and processors and one for stiff integration methods for large systems of particles.*

In addition, it is clear to the committee that the management, analysis, and mining of data present an increasingly critical and crosscutting algorithmic challenge. Enormous sets of input data, such as those from satellites and telescopes, require HECC to digest data and elicit insights.

Conclusion 7. *To capitalize on HECC's promise for overcoming the major challenges in many fields, there is a need for students in those fields, graduate and undergraduate, who can contribute to HECC-enabled research and for more researchers with strong skills in HECC.*

The committee foresees a growing need for computational scientists and engineers who can work with mathematicians and computer scientists to develop next-generation HECC software. Chapters 4, 5, and 6 explicitly mention a need for the more widespread teaching of scientific computing. Specifying an optimal career path for people who are able to straddle HECC and a traditional discipline is problematic, especially in academia. What is needed is a path that encompasses both a service role (HECC consulting within their field and to computer scientists) and opportunities to conduct their own research.

Even though the four fields selected for this study are disparate, the committee was able to develop major challenges for each and then determine which of those challenges are critically dependent on HECC. The following are suggestions for evaluating the potential impacts of HECC in other fields:

- It is necessary to build on the existing consensus about a field's current frontiers or major challenges. Developing from scratch a consensus picture of the frontier and of the major challenges that define promising directions for extending that frontier is in itself a sizable task.
- It is important to determine which major challenges for the field are critically dependent on HECC. While it is easy to spot opportunities for applying HECC to gain advantage, that is not the same as identifying the challenges whose progress will be impeded without the use of appropriate HECC.

All the infrastructure components needed to apply HECC to the challenges that depend on it must be identified, and the community must develop a clear understanding of the resources needed to complete the infrastructure. Merely giving a field access to supercomputers is no guarantee that the field's scientific progress will be enabled or accelerated.

1

Problem Definition and History

INTRODUCTION

This report identifies the major scientific and technical challenges in four fields of science and engineering that are critically dependent on high-end capability computing (HECC), and it characterizes the ways in which they depend on HECC. The committee thinks of its study as a gaps analysis, because it looked at the science and engineering "pull" on computing rather than the technology "push" enabled by computing. This perspective complements the more typical one, in which a new or envisioned computing capability arises and is followed by studies on how to profit from it.

It is generally accepted that computer modeling and simulation offer substantial opportunities for scientific breakthroughs that cannot otherwise—using laboratory experiments, observations, or traditional theoretical investigations—be realized. At many of the research frontiers discussed in this report, computational approaches are essential to continued progress and will play an integral and essential role in much of twenty-first century science and engineering. This is the inevitable result of decades of success in pushing those frontiers to the point where the next logical step is to characterize, model, and understand great complexity.

It became apparent during the committee's analysis of the charge, especially task (b), that "high-end capability computing" needs to be interpreted broadly. A 2005 report from The National Research Council (NRC, 2005) defines capability computing as "the use of the most powerful supercomputers to solve the largest and most demanding problems."[1] The implication is that capability computing expands the range of what is possible computationally. In computationally mature fields, expanding that range is generally accomplished by such steps as increasing processing power and memory and improving algorithmic efficiency. The emphasis is on the most powerful supercomputers. But in other fields, computationally solving the "largest and most demanding problems" can be limited by factors other than availability of extremely powerful supercomputers. In those situations, capability computing is still whatever expands the range of what is possible computationally, but it might not center around the use of "the most powerful" supercomputers.

---

[1] Quote from p. 279 of the NRC report.

Thus, in this report, capability computing is interpreted as computing that enables some new science or engineering capability—an insight or means of investigation that had not previously been available. In that context, high-end capability computing is distinguished by its ability to enable science and engineering investigations that would otherwise be infeasible.

HECC is also distinguished by its nonroutine nature. As the term "capability computing" is normally used, this corresponds to computing investments that are more costly and risky than somewhat time-tested "capacity computing." This report retains that distinction: HECC might require extra assistance—for example, to help overcome the frailties encountered with a first-time implementation of a new model or software—and entail more risk than more routine uses of computing. But the committee's focus on increasing scientific and engineering capabilities means that it must interpret "computing investments" to include whatever is needed to develop those nonroutine computational capabilities. The goal is to advance the fields. Progress might be indicated by some measure of computational capability, but that measure is not necessarily just processing power.

Targeted investments might be needed to stimulate the development of some of the components of HECC infrastructure, and additional resources might be necessary to give researchers the ability to push the state of the art of computing in their discipline. These investments would be for the development of mathematical models, algorithms, software, hardware, facilities, training, and support—any and all of the foundations for progress that are unlikely to develop optimally without such investments given the career incentives that prevail in academe and the private sector. Given that the federal government has accepted this responsibility (see the section below on history), it is faced with a policy question: How much HECC infrastructure is needed, and of what kind? To answer this, the committee sought and analyzed information that would give it two kinds of understanding:

1. An understanding of the mix of research topics that is desirable for the nation. Each field explicitly or implicitly determines this for itself through its review of competing research proposals and its sponsorship of forward-looking workshops and studies. Federal policy makers who define programs of research support and decide on funding levels are involved as well.
2. An understanding of the degree to which nationally important challenges in science and engineering can best be met through HECC. Answering the question about what kind of infrastructure is needed requires an understanding of a field's capacity for making use of HECC in practice, not just *potential* ways HECC could contribute to the field. In that spirit, this report considers "HECC infrastructure" very broadly, to encompass not just hardware and software but also training, incentives, support, and so on.

This report supplies information that is needed to gain the insights described in (1) and (2) above. It also suggests to policy makers a context for weighing the information and explains how to work through the issues for four disparate fields of science and engineering. The report does not, however, present enough detail to let policy makers compare the value to scientific progress of investments in HECC with that of investments in experimental or observational facilities. Such a comparison would require estimates of the cost of different options for meeting a particular research challenge (not across a field) and then weighing the likelihood that each option would bring the desired progress.

Computational science and engineering is a very broad subject, and this report cannot cover all of the factors that affect it. Among the topics not covered are the following, which the committee recognizes as important but which it could not address:

- The current computing capabilities available today for the four fields investigated and specific projections about the computations that would be enabled by a petascale machine.
- The need for computing resources beyond today's emerging capabilities and desirable features of future balanced high-end systems.
- Pros and cons associated with the use of community codes (although they are discussed briefly in Chapter 6).
- The policies of various U.S. government funding agencies with respect to HECC.
- The ability of the academic community to build and manage computational infrastructure.
- Policies for archiving and storing data.

HISTORY OF HIGH-END COMPUTING

The federal government has been a prime supporter of science and engineering research in the United States since the 1940s. Over the subsequent decades, it established a number of federal laboratories (primarily oriented toward specific government missions) and many intramural and extramural research programs, including an extensive system for supporting basic research in academia for the common good. Until the middle decades of the twentieth century, most of this research could be classified as either theoretical or experimental.

By the 1960s, as digital computing evolved and matured, it became widely appreciated that computational approaches to scientific discovery would become a third mode of inquiry. That idea had, of course, already been held for a number of years—at least as early as L.F. Richardson's experiment with numerical weather prediction in 1922 (Richardson, 1922) and certainly with the use of the ENIAC in the 1940s for performing ballistics calculations (Goldstine and Goldstine, 1946). By the 1970s, the confluence of computing power, robust mathematical algorithms, skilled users, and adequate resources enabled computational science and engineering to begin contributing more broadly to research progress (see, for instance, Lax, 1982).

In its role of furthering science and engineering for the national interest, the federal government has long accepted the responsibility for supporting high-end computing, beginning with the ENIAC. Clearly, the ENIAC would not be considered a supercomputer today (nor would the Cray-1, to choose a cutting-edge technology from the late 1970s), but high-end computing is commonly defined as whatever caliber of computing is pushing the state of art of computing at any given time. Similarly, today's teraflop[2] computing is becoming fairly routine within the supercomputing community, and some would no longer consider computing at a few teraflops as being at the high end. Petascale computing will be the next step,[3] and some are beginning to think about exascale computing, which would represent a further thousandfold increase in capability beyond the petascale.

Science and engineering progress over many years has been accompanied by the development of new tools for examining natural phenomena. The invention of microscopes and telescopes four centuries ago enabled great progress in observational capabilities, and the resulting observations have altered our views of nature in profound ways. Much more recently, techniques such as neutron scattering, atomic force microscopy, and others have been built on a base of theory to enable investigations that were otherwise impossible. The fact that theory underpins these tools is key: Scientists needed a good

---

[2] The prefix "tera-" connotes a trillion, and "flop" is an acronym for "floating point operations."

[3] Los Alamos National Laboratory announced in June 2008 that it had achieved processing speeds of over 1 petaflop/s for one type of calculation; see the news release at http://www.lanl.gov/news/index.php/fuseaction/home.story/story_id/13602. Accessed July 18, 2008.

understanding of (in the cases just cited) subatomic nature before they could even imagine such probes, let alone engineer them.

Computational tools are analogous to observational tools. Computational simulation enables us to explore natural or man-made processes that might be impossible to sense directly, be just a hypothetical creation on a drawing board, or be too complex to observe in adequate detail in nature. Simulations are enabled by a base of theory sufficient for creating mathematical models of the system under study, which also provides a good understanding of the limitations of those models. Richardson's experiment in weather prediction could not have been developed without a mathematical model of fluid flow, and Newton's laws were necessary in order to run ballistics calculations on the ENIAC. But the theoretical needs are much deeper than just the understanding that underpins mathematical models. For instance, Richardson's projections did not converge because the necessary numerical analysis did not yet exist, and so the approximation used was inappropriate for the task in ways that were not then understood. More recently (since the 1970s), the development of ever more efficient algorithms for the discretization of differential equations and for the solution of the consequent linear algebra formulations has been essential to the successes of computational science and engineering. In fact, it is generally agreed that algorithmic advances have contributed at least as much as hardware advances to the increasing capabilities of computation over the past four decades.

Historically, there has been a strong coupling between the development of algorithms and software for scientific computing and the fundamental mathematical understanding of the underlying models. For example, beginning in the 1940s mathematicians such as von Neumann, Lax, and Richtmyer (see, for example, von Neumann and Goldstine, 1947; Lax and Richtmyer, 1956) were deeply involved in investigating the well-posedness properties of nonlinear hyperbolic conservation laws in the presence of discontinuities such as shock waves. At the same time they were also developing new algorithmic concepts to represent such discontinuous solutions numerically. The resulting methods were then incorporated into simulation codes by the national laboratories and industry, often in collaboration with the same mathematicians. This interaction between numerical algorithm design and mathematical theory for the underlying partial differential equations has continued since that time, leading to methods that make up the current state of the art in computational fluid dynamics today: high-resolution methods for hyperbolic conservation laws; projection methods and artificial compressibility methods for low-Mach-number fluid flows; adaptive mesh refinement methods; and a variety of methods for representing sharp fronts.

A similar connection can be seen in the development of computer infrastructure and computational science. The first computers were developed for solving science and engineering problems, and the earliest development efforts in a variety of software areas from operating systems to languages and compilers were undertaken to make these early computers more usable by scientists. Over the last 30 years, mathematical software such as LINPACK and its successors for solving linear systems have become standard benchmarks for measuring the performance of new computer systems. One strain of current thinking in the area of computer architecture is moving toward the idea that the algorithm is the fundamental unit of computation around which computer performance should be designed, with most of the standard algorithms for scientific applications helping to define the design space (Asanovic et al., 2006).

A lot of the technology for doing computing (e.g., for computational fluid dynamics) was subsidized by national security enterprises from the 1960s through the 1980s. To a large extent, those enterprises have become more mission focused, and so we can no longer assume that all the components of future HECC infrastructure for science and engineering generally, such as algorithm development and visualization software, will be created in federal laboratories. This heightens the need to examine the entire

HECC infrastructure and consciously determine what will be needed, what will be the impediments, and who will be responsible.

CURRENT STATE OF HIGH-END CAPABILITY COMPUTING

By thinking of HECC as whatever level of computing capability is nonroutine, it follows that the definition of capability computing is inextricably bound to the capabilities of a given field and a particular science or engineering investigation. Computational science and engineering is a systems process, bringing together hardware, software, investigators, data, and other components of infrastructure to produce insight into some question. Capability computing can be nonroutine due to limitations in any of these infrastructure components. Many factors constrain our ability to accomplish the necessary research, not just the availability of processing cycles with adequate power. Emphasizing one component over the others, such as providing hardware alone, does not serve the real needs of computational science and engineering. For instance, one recent HECC fluid dynamics simulation produced 74 million files; just listing all of them would crash the computer. A systems approach to HECC will consider these sorts of practical constraints as well as the more familiar constraints of algorithms and processing speed. Some scientific questions are not even posed in a way that can map onto HECC, and creating that mapping might be seen as part of the HECC infrastructure.

Typically, simulations that require HECC have complications that preclude an investigator simply running them for a long time on a dedicated workstation. Overcoming those complications often requires nonroutine knowledge of a broad range of disciplines, including computing, algorithms, data management, visualization, and so on. Therefore, because the individual investigator model might not suffice for HECC, mechanisms that enable teamwork are an infrastructure requirement for such work. This is consistent with the general need in computational science and engineering to assemble multidisciplinary teams.

While a number of federal agencies provide high-end hardware for general use, other aspects of HECC infrastructure are handled in a somewhat ad hoc fashion. For instance, middleware (e.g., data management software) and visualization interfaces might be created in response to specific project requirements, with incomplete testing and documentation to allow them to truly serve as general-purpose tools, and the creators might be graduate students who then move to other facilities. Data repositories and software might not be maintained beyond the lifetime of the particular research grant that supported their development. "Hardening" of software, development of community codes, and other common-good tasks might not be done by anyone. In short, many of the components of HECC infrastructure are cobbled together, resulting in a mix of funding horizons and purposes. While the DOE national laboratories, in particular, have succeeded in developing environments that cover all the components of HECC infrastructure, more often the systems view is lacking. Thus, the United States may not have the best environments for enabling computational advances in support of the most pressing science and engineering problems.

This report does not prognosticate about future computing capabilities, a task that would be beyond the study's charge and the expertise of the committee.[4] The committee relies on the following vision of the next generation (NSF, 2006, p. 13):

> By 2010, the petascale computing environment available to the academic science and engineering community is likely to consist of: (i) a significant number of systems with peak performance in the 1-50 tera-

---

[4]NRC (2005) is a good source for exploring this topic.

flops range, deployed and supported at the local level by individual campuses and other research organizations; (ii) multiple systems with peak performance of 100+ teraflops that support the work of thousands of researchers nationally; and, (iii) at least one system in the 1-10 petaflops range that supports a more limited number of projects demanding the highest levels of computing performance. All NSF-deployed systems will be appropriately balanced and will include core computational hardware, local storage of sufficient capacity, and appropriate data analysis and visualization capabilities.

HECC challenges abound in science and engineering, and the focus of this report on four fields should not be taken to imply that those particular fields are in some sense special. The committee is well aware of the important and challenging opportunities afforded by HECC in many other fields.

REFERENCES

Asanovic, K., Ras Bodik, Bryan Christopher Catanzaro, et al. 2006. *The Landscape of Parallel Computing Research: A View from Berkeley*. Technical Report No. UCB/EECS-2006-183. Available at http://www.eecs.berkeley.edu/Pubs/TechRpts/2006/EECS-2006-183.pdf. Accessed July 18, 2008.

Goldstine, H.H., and A. Goldstine. 1946. *The Electronic Numerical Integrator and Computer (ENIAC)*. Reprinted in *The Origins of Digital Computers: Selected Papers*. New York, N.Y.: Springer-Verlag, 1982, pp. 359-373.

Lax, Peter (ed.). 1982. "Large Scale Computing in Science and Engineering." National Science Board.

Lax, P.D., and R.D. Richtmyer. 1956. Survey of the stability of linear finite difference equations. *Communications on Pure & Applied Mathematics* 9: 267-293.

NRC (National Research Council). 2005. *Getting Up to Speed: The Future of Supercomputing*. Washington, D.C.: The National Academies Press.

NSF (National Science Foundation). 2006. *NSF's Cyberinfrastructure Vision for 21st Century Discovery*. Draft Version 7.1. Available at http://www.nsf.gov/od/oci/ci-v7.pdf. Accessed July 18, 2008.

Richardson, Lewis Fry. 1922. *Weather Prediction by Numerical Process*. Cambridge, England: Cambridge University Press.

von Neumann, J., and H.H. Goldstine. 1947. Numerical inverting of matrices of high order I. *Bulletin of the American Mathematical Society* 53(11): 1021-1099.

2

The Potential Impact of HECC in Astrophysics

INTRODUCTION

It is impossible to gaze at the sky on a clear, moonless night and not wonder about our Universe and our place in it. The same questions have been posed by mankind for millennia: When and how did the Universe begin? What are the stars and how do they shine? Are there other worlds in the Universe like Earth? The answers to many of these questions are provided by scientific inquiry in the domains of astronomy and astrophysics.

Astronomy is arguably the oldest of all the sciences. For most of history, observations were made with the naked eye and therefore confined to the visible wavelengths of light. Today, astronomers use a battery of telescopes around the world and on satellites in space to collect data across the entire electromagnetic spectrum—that is, radio, infrared, optical, ultraviolet, X-ray, and gamma-ray radiation.

Astrophysics emerged as a discipline in the last century as scientists began to apply known physical laws to interpret the structure, formation, and evolution of the planets, stars, and galaxies, and indeed the Universe itself. Astronomical systems have long challenged our understanding of physics; for example, research in nuclear physics is intimately tied to our desire to understand fusion reactions inside stars. Similarly, some of the most important discoveries made by astronomical observers have resulted from the predictions of theoretical physics, such as the cosmic microwave background radiation that is a signature of the Big Bang.

The rate of discovery in astronomy and astrophysics is rapid and accelerating. It is only within the past 100 years that we have understood the size and shape of our own Milky Way galaxy, and that there are billions of other galaxies like the Milky Way in the Universe. It is only in the past 50 years that we have realized the Universe had a beginning in what is now called the Big Bang. It is only within the past decade that we have narrowed the uncertainties surrounding the age of the Universe. Some of the most important discoveries made in the past decade are these three:

- *Dark matter and dark energy.* It is long been known that the motions of galaxies in clusters, and their internal rates of rotation, require the existence of dark matter—that is, material in the

Universe that interacts gravitationally but not with electromagnetic radiation. Precise measurements of the expansion history of the Universe, made using distances determined from Type Ia supernovae, have recently indicated the existence of dark energy, which is responsible for an acceleration with time in the expansion rate of the Universe that counteracts the deceleration produced by gravity. In the past year, the Wilkinson Microwave Anisotropy Probe (WMAP), a satellite launched by NASA in 2001, has provided the most accurate observations of the fluctuations in the cosmic microwave background radiation that were first discovered by the Cosmic Background Explorer (COBE), another NASA mission, launched in 1989. These fluctuations are the imprint of structure in the Universe that existed about 300,000 years after the Big Bang, structure that eventually collapsed into the galaxies, stars, and planets we see today. These observations not only challenge astrophysicists to explain how these fluctuations grew and evolved into the structure we see today, but they also provide firm evidence that verifies the reality of both dark matter and dark energy. Discovering the nature of dark matter and dark energy are perhaps the two most compelling challenges that face astrophysicists today.

- *Exosolar planets.* In the past 10 years, over 200 new planets have been discovered orbiting stars other than the Sun. Most of these planets are dramatically different from those in our solar system. For example, a significant fraction consist of large, Jupiter-like bodies that orbit only a few stellar radii from their host stars. There is no analog to such planets in our own solar system. Indeed, our current theory of planet formation predicts that Jupiter-like planets can form only in the outer regions of planetary systems, at distances several times the Earth-Sun separation. Thus, the existence of these new planetary systems challenges our understanding of planet formation and the dynamical evolution of planetary systems. As our techniques for detecting exosolar planets (also known as exoplanets) improve, astronomers are finding more and more planets closer in size to and with properties similar to those of Earth. (As of the time of this report, several planets with masses only a few times that of Earth have been found.) Of course, this raises the question of whether any of them harbor life.

- *Unambiguous evidence for the existence of black holes.* It has long been known that there is a compact object at the center of our galaxy responsible for producing powerful nonthermal emission at a variety of wavelengths. One way to delimit the nature of this object is to measure its size and mass: If the inferred mass density is sufficiently high, then it must be a black hole. Using diffraction-limited images in the infrared to measure the motions of stars within 1 light-year of this object over the past 10 years, two research groups have been able to limit its mass to at least 3 million solar masses. There is no known object in the Universe, other than a supermassive black hole, that could contain this much mass in such a small volume. Similar observations of the motions of stars and gas near the centers of several other galaxies, notably NGC4258, also known as Messier106, have also provided unambiguous evidence for the existence of supermassive black holes.

The purpose of this chapter is to assess the potential impact of HECC on the progress of research in astrophysics. Using the discoveries of the past decade, along with sources such as prior decadal surveys of the field and input from outside experts, the committee formulated a list of the major challenges facing astrophysics today, identified the subset of challenges that require computation, and investigated the current state of the art and the future impact of HECC on progress in facing these challenges.

MAJOR CHALLENGES IN ASTROPHYSICS

Several reports published recently in consultation with the entire astronomy and astrophysics community have identified key questions that confront the discipline in the coming decade. These made the task of identifying the major challenges in astrophysics much simpler for the committee. In particular, the NRC decadal survey of astronomy and astrophysics, *Astronomy and Astrophysics in the New Millennium*, published in 2001 (often referred to as the McKee-Taylor report), the NRC report *Connecting Quarks with the Cosmos: Eleven Science Questions for the New Century*, published in 2002 (also called the Turner report), and the National Science and Technology Council (NSTC) report *The Physics of the Universe: A Strategic Plan for Federal Research at the Intersection of Physics and Astronomy*, published in 2004, were instrumental in developing the list of questions summarized in this section. At its first meeting, the committee heard a presentation from Chris McKee, coauthor of the McKee-Taylor report, and at a later meeting it heard presentations from Tom Abel, Eve Ostriker, Ed Seidel, and Alex Szalay on topics related to the identification of the major challenges and the potential impact of HECC on them.

Committee members found themselves in complete agreement with the consistent set of major challenges identified in each of these three published reports. The challenges take the form of questions that are driving astrophysics and that are compelling because our current state of knowledge appears to make the challenges amenable to attack:

1. What is dark matter?
2. What is the nature of dark energy?
3. How did galaxies, quasars, and supermassive black holes form from the initial conditions in the early Universe observed by WMAP and COBE, and how have they evolved since then?
4. How do stars and planets form, and how do they evolve?
5. What is the mechanism for supernovae and gamma-ray bursts, the most energetic events in the known Universe?
6. Can we predict what the Universe will look like when observed in gravitational waves?

Observations, Experiment, Theory, and Computation in Astrophysics

As in other fields of science, astrophysicists adopt four modes of investigation: observation, experiment, theory, and computation. Astronomy is characterized by its reliance on observation over experimentation, and this clearly affects the information available to astrophysicists. Virtually all that we know about the Universe beyond the solar system comes from electromagnetic radiation detected on Earth and in space. To push the frontiers of our knowledge, astronomers build ever-larger telescopes that operate over ever-wider bands of the electromagnetic spectrum and equip them with more efficient and more sensitive digital detectors and spectrographs. This is leading to an explosion of data in digital form, a point to which we return below.

Experimentation has a long and distinguished history in astrophysics. Although it is of course impossible to build a star in the laboratory and perform experiments on it, it is possible to measure basic physical processes important in stars and other astrophysical systems in the laboratory. For example, the cross sections for nuclear reactions of relevance to astrophysics have been the subject of laboratory measurements for many decades, as have the cross sections for the interaction of astrophysically abundant ions with light. More recently, the construction of high-energy-density laser and plasma fusion devices have enabled experiments on the dynamics of plasmas at the pressures and temperatures relevant to a variety of astrophysical systems.

Theory is primarily concerned with the application of known physical laws to develop a mathematical model of astrophysical phenomena. The goal is to identify the most important physical effects and to formulate the simplest mathematical model that adequately describes them. For example, the equations of stellar structure, which can be used to describe the evolution of stars, result from combining mathematical models of processes in nuclear physics, gas dynamics, and radiation transfer. Comparing the predictions of such a model to observations allows investigators to develop insight about the details of those component processes and to infer the degree to which those processes contribute to and account for known data on stellar evolution.

Finally, computation has emerged as a powerful means to find solutions to mathematical formulations that are too complex to solve analytically. Computation also allows "numerical experimentation"—that is, systematic exploration of the effects of varying parameters in a mathematical model. This is particularly important in astrophysics, because so many phenomena of interest cannot be replicated in the laboratory. Finally, data analysis and modeling are critical in observational astronomy, where it is important to find interesting and unusual objects out of billions of candidates (for example, the most distant galaxies) or where information can be extracted only from a statistical model of very large data sets (for example, patterns in the distribution of galaxies in the Universe that reveal clues to the nature of dark matter and dark energy). Computation has a long and distinguished history in astrophysics. The development of many numerical algorithms has been driven by the need to solve problems in astrophysics. Similarly, astrophysicists have relied heavily on HECC since the first electronic computers became available in the 1940s.

To identify those challenges that are critically reliant on HECC, the committee considered the importance of each of these four modes of investigation to the solution of the six major challenges listed on the preceding page. The sections below discuss in more detail the major challenges in astrophysics that require HECC, while also describing the current state of the art in algorithms and hardware and what is needed to make substantial progress.

MAJOR CHALLENGES THAT REQUIRE HECC

It seems likely that the bulk of the progress on Major Challenges 1 and 2 will come from more advanced astronomical observations. Even then, it is important to emphasize that some of the data sets required to inform our understanding of dark matter and dark energy are unprecedented in size and will require HECC for their analysis. The need for HECC to exploit large data sets is discussed on its own later on, in a separate section.

For the remaining four challenges, the committee concluded that while further observation will also play a role in answering the questions, it is unlikely that any significant advance in understanding will be achieved without a major, if not dominant, reliance on HECC. That so many of the major challenges in astrophysics require HECC should not be surprising given the complexity of astrophysical systems, the need for numerical experimentation as proxies for laboratory experiments, and the long tradition of the use of HECC in astrophysics, which suggests that the field is comfortable with cutting-edge computation.

Major Challenge 3. Understanding the Formation and Evolution of Galaxies, Quasars, and Supermassive Black Holes

Precise measurements of anisotropies in the cosmic microwave background over the past decade have reduced uncertainties in the fundamental cosmological parameters to a few percent or less and have provided a standard model for the overall properties of the Universe. According to this model, the Universe is geometrically flat and consists presently of about 4 percent ordinary matter, 22 percent nonbaryonic dark matter, and 74 percent dark energy. It is thought that small-amplitude density fluctuations, initially seeded at early times by a process like inflation, grew through gravitational instability to produce stars, galaxies, and larger-scale structures over billions of years of evolution.

Thus cosmic microwave background measurements provide the initial conditions we need, in principle, to understand the structure of the present-day Universe. However, while the gravitational dynamics of dark matter are reasonably well understood, the physics attending the formation of stars and galaxies is highly complex and involves an interplay between gravity, hydrodynamics, and radiation. Initially smooth, material in slightly overdense regions collapsed, shock heating the gas and leaving behind halos of dark and ordinary matter. In some cases, the baryons in these halos cooled radiatively, producing the dense, cold gas needed to form stars. Winds, outflows, and radiation from stars, black holes, and galaxies established a feedback loop, modifying the intergalactic medium and influencing the formation of subsequent generations of objects. Numerical simulation makes it possible to follow the coupled evolution of dark matter, dark energy, baryons, and radiation so that the physics of this process can be inferred and the properties of the Universe can be predicted at future epochs.

It is instructive to consider the current state of the art in the mathematical models and numerical algorithms for simulating galaxy formation. As indicated above, most of the mass in the Universe is in the form of dark matter, which in the simplest interpretation interacts only gravitationally and is collisionless. The equations describing the evolution of this component are just Newton's laws (or Einstein's equations on larger scales) supplemented by Poisson's equation for the gravitational interaction of the dark matter with itself or with baryons. Stars in galaxies also interact gravitationally and can be approximated as a collisionless fluid except in localized regions, where the relaxation time may be relatively short. The gas that cooled and formed stars and galaxies evolved according to the equations of compressible hydrodynamics, and this material evolved under the influence of gravity (its own self-gravity, as well as that arising from dark matter and stars), pressure gradients, shocks, magnetic fields, cosmic rays, and radiation. Because the gas can cool and be heated by radiation, the equations of radiative transfer also must ultimately be solved along with the equations of motion for stars, gas, and the inferred behavior of dark matter. But to date there have been few attempts to dynamically couple radiation and gas owing to the complexity of the physics and the extreme numerical resolution needed.

Various numerical algorithms have been implemented to solve the equations underlying galaxy formation. Dark matter and stars are invariably represented using N-body particles because it is impractical to solve Boltzmann's equation as a six-dimensional partial differential equation. The most computationally challenging aspect of the simulations is solving for the gravitational field in an efficient fashion with a dynamic range in spatial scales sufficiently large to capture the distribution of galaxies on cosmological scales as well as to describe their internal structure. State-of-the-art codes for performing these calculations typically employ a hybrid scheme for solving Poisson's equation. For example, in the TreePM approach, the gravitational field is split into short- and long-range contributions. Short-range forces are computed with a tree method, in which particles are grouped hierarchically and their contribution to the potential is approximated using multipole expansions. Long-range forces are determined using a particle-mesh technique by solving Poisson's equation using fast Fourier transforms. An advantage of this hybrid formulation is that long-range forces need to be updated only infrequently.

This approach has been used to perform very large N-body simulations describing the evolution of dark matter in the Universe, ignoring the gas. For example, the recent Millennium simulation (Springel et al., 2005) employed more than 10 billion particles to study the formation of structure in a dark-matter-only Universe, making it possible to quantify the mass spectrum of dark-matter halos and how this evolves with time. An example of such a calculation, in a smaller volume with fewer particles, is shown in Figure 2-1.

In fact, the problem of understanding galaxy formation is even more complex than solving a coupled set of differential equations for dark matter, stars, gas, and radiation, because some of these components can transform into one another. Stars, as well as supermassive black holes, form from interstellar gas, converting material from collisional to collisionless form. As stars evolve and die, they recycle gas into the interstellar medium and intergalactic medium, enriching these with heavy elements formed through nuclear fusion. It is also believed that radiative energy produced during the formation of supermassive black holes, possibly during brief but violent quasar phases of galaxy evolution, can impart energy and/or momentum into surrounding gas, expelling it from galaxies and heating material in the intergalactic medium.

Star formation and the growth of supermassive black holes are not well-understood, and they occur on scales that cannot be resolved in cosmological simulations. But since these processes are essential ingredients in the formation and evolution of galaxies, simplified descriptions of them are included in the mathematical model of galaxy formation, usually incorporated as subgrid-scale functions. An important goal for the field in the coming decades will be to formulate from first principles physically motivated models of star formation and black hole growth that can be incorporated into cosmological simulations.

Because current simulations are limited, many aspects of galaxy formation and evolution are not well understood. For example, in the local Universe, the vast majority of galaxies are classified as spirals.

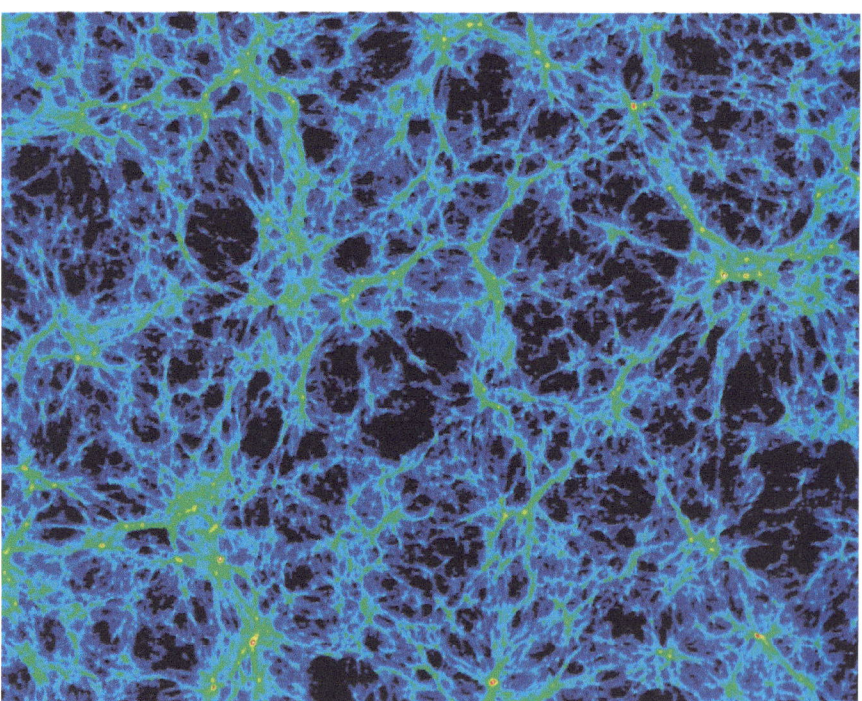

FIGURE 2-1 Snapshot from a pure *N*-body simulation showing the distribution of dark matter at the present time (light colors represent the greater density of dark matter). This square is the projection of a slice through a simulation box that is 150 megaparsecs (Mpc) across. One billion particles were employed.

Similar to the Milky Way and unlike elliptical galaxies, which are spheroidal and are supported against gravity by the random motions of their stars, spiral galaxies have most of their mass in thin, rotationally supported disks. It is believed that most (perhaps all) galaxies were born as spirals and that some were later transformed into ellipticals by collisions and mergers between them, a process that is observed to occur often enough to explain the relative abundances of the two types of galaxies. Indeed, simulations of mergers between individual, already-formed spirals show that the remnants are remarkably similar to actual ellipticals, strengthening the case for this hypothesis.

Yet in spite of the obvious significance of spiral galaxies to our picture of galaxy evolution, simulations of galaxy formation starting from cosmological initial conditions have invariably failed to produce objects with structural and kinematic properties resembling those of observed galactic disks. The cause for this failure is presently unknown but probably involves an interplay between inadequate resolution and a poor representation of the physics describing star formation and associated feedback mechanisms.

Cosmological simulations that incorporate gas dynamics, radiative processes, and subgrid models for star formation and black-hole growth have employed a variety of algorithms for solving the relevant dynamical equations. In applications to cosmology and galaxy formation, state-of-the-art simulations are being done solving the equations of hydrodynamics using either a particle-based approach (smoothed-particle hydrodynamics, or SPH) or a finite-difference solution on a mesh. The most promising grid-based schemes usually employ some variant of adaptive mesh refinement (AMR), which makes it possible to achieve a much larger range in spatial scales than would be possible with a fixed grid. Radiative transfer effects have typically not been handled in a fully self-consistent manner but have instead been included as a postprocessing step, ignoring the dynamical coupling to the gas. An example of such a calculation is illustrated in Figure 2-2, which shows the impact of ionizing radiation from an assumed distribution of galaxies in the simulation shown in Figure 2-1.

These various methods have already yielded a number of successes applied to problems in cosmology and galaxy formation. For example, simulations of the structure of the intergalactic medium have provided a new model for the absorption lines that make up the Lyman-alpha forest seen in the spectra of high-redshift galaxies, according to which the absorbing structures are the filaments seen in, for example, Figures 2-1 and 2-2 (the "Cosmic Web"; for a review, see Faucher-Giguere et al., 2008). High-resolution N-body simulations have quantified the growth of structure in the dark matter and the mass function of halos. Multiscale calculations have plausibly shown that the feedback loop described above probably plays a key role in determining the properties of galaxies and the hosts of quasars.

However, in the future, simulations with even greater dynamic range and physical complexity will be needed for interpreting data from facilities such as the Giant Magellan Telescope, the Thirty-Meter Telescope, the James Webb Space Telescope, and the Square Kilometer Array, which will enable us to probe the state of the Universe when stars first began to form, ending the Cosmic Dark Ages. For example, the self-consistent interaction between the hydrodynamics of the gas and the evolution of the radiation field must be accounted for to properly describe the formation of the first objects in the Universe and the reorganization of the intergalactic medium. Large galaxies forming later were shaped by poorly understood processes operating on smaller scales than those characterizing the global structure of galaxies—such as, in particular, star formation, the growth of supermassive black holes, and related feedback effects—and the physical state of the interstellar medium within galaxies, which is also poorly understood. Understanding the formation and evolution of galaxies and their impact on the intergalactic medium is at the forefront of research in modern astrophysics and will remain so for the foreseeable future. The physical complexity underlying these related phenomena cannot be described using analytic methods alone, and the simulations needed to model them will drive the need for new mathematical models, sophisticated algorithms, and increasingly powerful computing resources.

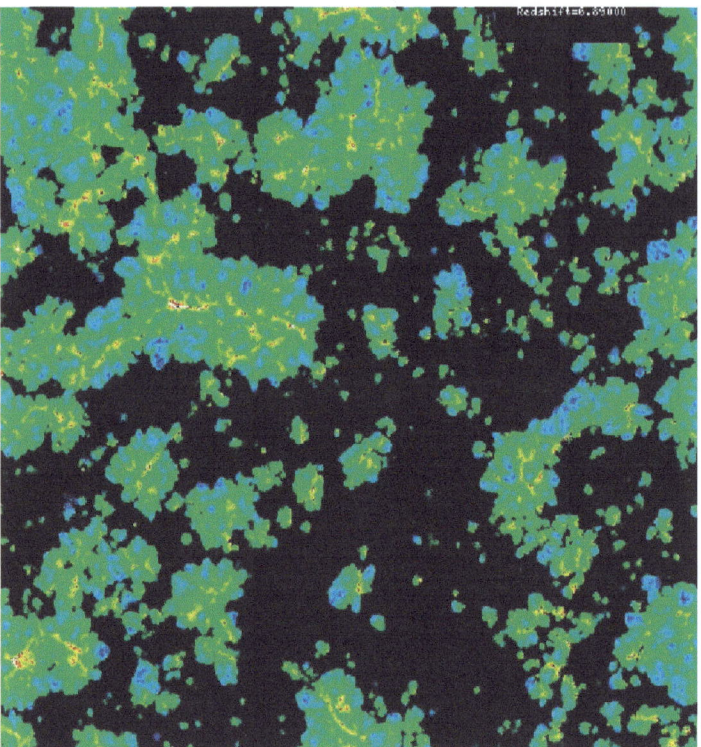

FIGURE 2-2 Simulation from Figure 2-1 post-processed to demonstrate the impact of ionizing radiation from galaxies. Black regions are neutral, while colored regions are ionized. The ionizing sources are in red.

It is equally instructive to consider what scale of computation is required, in terms of both the mathematical models and the numerical algorithms, to achieve a revolutionary step forward in our understanding of galaxy formation. The challenge is associated with the vast range of spatial scales that must be covered. A volume of the local Universe containing a representative sample of galaxies will be on the order of a few hundred megaparsecs across, if not larger. Star-forming events within galaxies occur on the scale of a few parsecs or less, implying a linear dynamic range greater than 10^8. Even smaller scales (about 0.01 parsec or less) could be important if processes related to the growth of supermassive black holes in the centers of galaxies influence the dynamical state of nearby star-forming gas. If this problem were to be attacked using, say, a uniform grid, the number of cells required in three dimensions would be between 10^{24} and 10^{30}, whereas we can currently handle only between 10^{10} and 10^{12} cells. More plausibly, advances in the near term will probably be made using codes with adaptive resolution, integrating different algorithms in a multiscale manner, so that the entire range of scales need not be captured in a single calculation. For example, a global simulation of a galaxy or a pair of interacting galaxies could be grafted together with a separate code following the inflow of gas onto supermassive black holes to estimate the impact of radiative heating and radiation pressure on the gas and dust in the vicinity of the black holes, to study the impact of black hole feedback on galaxy evolution.

The algorithms that are required would, ideally, solve the equations of radiation hydrodynamics, including gravitational interactions between gas and collisionless matter, and be capable of handling magnetohydrodynamical effects. This would almost certainly involve a combination of adaptive mesh and particle codes that can represent the various components in the most efficient manner possible.

The algorithms would likely be running across a number of computing architectures simultaneously to maximize throughput, depending on the scalability of different aspects of the calculation. In addition to needing large numbers of fast processors operating in parallel, simulations such as these will place stringent demands on core memory and disk storage. For example, a pure N-body simulation with $(10,000)^3$ particles, using the most efficient codes currently available, would require on the order of 100 Tbytes of core memory and would produce data at the rate of roughly 30 Tbyte per system snapshot, or on the order of 1,000 Tbytes total for relatively coarse time sampling. A simulation of this type would make it possible to identify all sources of ionizing radiation in volumes much larger than that in, for example, Figure 2-2 and hence enable theoretical predictions for the 21-cm emission from neutral gas that would be detectable by the Square Kilometer Array.

Major Challenge 4. Understanding the Formation and Evolution of Stars and Planets

All of the stars visible to the naked eye are in the Milky Way galaxy, and they formed long after the Big Bang. Roughly a century ago, astronomers realized that the Milky Way contains not just stars but also giant clouds of molecular gas and small solid particles, which they called dust. Further observations have revealed that new generations of stars form from the gravitational collapse and fragmentation of these Giant Molecular Clouds (GMCs). The formation process occurs in two stages. At first, dynamical collapse creates a dense, swollen object, called a proto-star, surrounded by a rotating disk of gas and dust. As the core cools, it shrinks, eventually evolving into a star once the central densities in the core become high enough to initiate nuclear fusion. During the latter phase, which can take millions of years for stars of our Sun's mass, it is thought that planets can form from the surrounding protostellar disk. Thus, the formation of stars is inextricably linked with that of planets.

Star Formation

Although astrophysicists now understand the basic physical processes that control star formation, a predictive theory that can explain the observed star formation rates and efficiencies in different environments has yet to emerge. The most basic physics that must be incorporated is the gas dynamics of the interstellar material, including the stresses imposed by magnetic fields. In addition, calculating the cooling and heating of the gas due to the emission and absorption of photons is both important and challenging. Radiation transfer is an inherently nonlocal process: Photons emitted in one place can be absorbed somewhere else in the GMC. Moreover, most of the cooling is through emission at particular discrete frequencies, and calculating such radiative transfer in a moving medium is notoriously complex. Finally, additional microphysical processes such as cosmic-ray ionization, recombination on grains, chemistry (which can alter the abundances of chemical species that regulate cooling), and diffusion of ions and neutrals can all affect the dynamics and must be modeled and incorporated. All of this physics occurs in the deeply nonlinear regime of highly turbulent flows, making analytical solutions impossible. Thus the vast majority of efforts to address the major challenges associated with star formation are based on computation.

It is instructive to consider the current state of the art in mathematical models and numerical algorithms for star and planet formation. Currently, the dynamics of GMCs are modeled using the equations of hydrodynamics or, more realistically, the equations of magnetohydrodynamics, including self-gravity and optically thin radiative cooling. Typical simulations begin with a turbulent, self-gravitating cloud, and they follow the collapse and fragmentation of the cloud into stars. A few simulations have considered the effects of radiation feedback using an approximate method (flux-limited diffusion) for radiative transfer, but these calculations require other simplifications (because they are not magnetohydrodynamical) to make them tractable. Either static or adaptive mesh refinement is a very powerful technique for resolving the collapse to small scales; however, robust methods for magnetohydrodynamics and radiation transfer on adaptively refined grids are still under active development. Typical calculations involve grids with a resolution of up to 1024^3 evolved for tens of dynamical times on the largest HECC platforms available today, using hundreds to thousands of processors. Figure 2-3 shows the results of one such calculation.

A typical AMR calculation uses 128^3 grids with up to seven levels of refinement (see, for example, Krumholz et al., 2007). Alternatively, some researchers use the smoothed particle hydrodynamics (SPH) algorithm to follow the collapse and fragmentation of GMCs; a typical calculation using about 10^6 particles will require several months of cpu time on a cluster with hundreds of processors (Bate and Bonnell, 2004). To date, there are no calculations that follow the chemistry and ionization of the gas self-consistently with the hydrodynamics.

As a result of such calculations, an entirely new paradigm for star formation is emerging (Heitsch et al., 2001). Previously, it was thought to be controlled by the slow, quasistatic contraction of a gravitationally bound core supported by magnetic pressure. These new numerical results indicate star formation is far more dynamic. Turbulence in the cloud generates transient, large-amplitude density fluctuations, some of which are gravitationally bound and collapse on a free-fall timescale (see Figure 2-3). Feedback from newly forming stars disperses the cloud and limits the efficiency of star formation. Many of the properties of newly formed stars are now thought to be determined by the properties of the GMCs from which they form.

It must be emphasized that the current simulations still lack important physics. For instance, there are no magnetohydrodynamical calculations of collapse using AMR. None of the calculations include a realistic treatment of the ionization and recombination processes that determine the degree of coupling to the magnetic field. No calculations yet include the galactic gravitational potential, including shocks

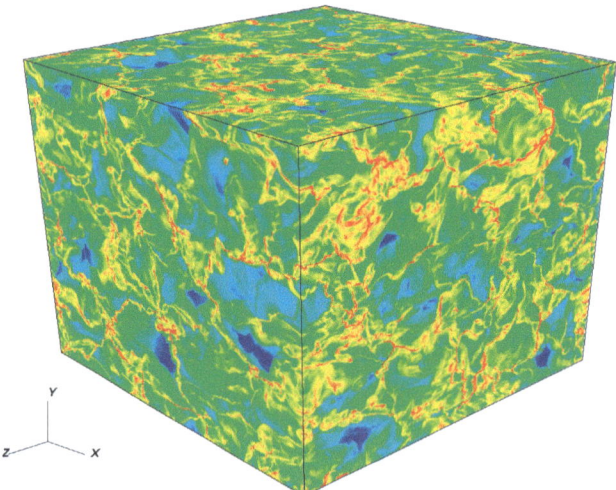

FIGURE 2-3 Structure of supersonic turbulence in a star-forming GMC, as revealed through three-dimensional hydrodynamic simulations on a 512^3 grid. Colors represent the gas density on the faces of the computational volume (red is highest density, blue lowest). A complex network of interacting shocks in the turbulence generates large density fluctuations, which can collapse to form stars.

induced by spiral arms, to follow the formation of GMCs from more diffuse phases of the interstellar medium. All of this physics is beyond current capabilities. Several outstanding problems in star formation could be addressed with such calculations, including the following:

- What controls the initial mass function—that is, the numbers of stars formed at different masses?
- What determines the efficiency of star formation—that is, the fraction of gas in a GMC that ultimately is turned into stars?
- By what process are GMCs formed in galaxies in the first place?
- How are planets formed in the accretion disks that surround newly forming stars? (This is discussed further below.)

Calculations that can address these issues are vital in view of recent observational programs that have shed new light on star formation. For example, the Spitzer Space Telescope is a billion-dollar NASA mission that has launched into space the most sensitive infrared telescope ever built. The mission is now returning unprecedented images and spectra of cold interstellar gas in the Milky Way and other galaxies. Interpreting the data from Spitzer requires more sophisticated theoretical and computational studies of star formation.

To achieve a revolutionary step forward in our understanding of star formation, future calculations need to consider nonideal magnetohydrodynamics in a partially ionized gas, including radiation transfer, self-gravity, ionization and recombination, and cosmic-ray transport. Ideally, the computational domain should include the entire galactic disk, so that the formation of GMCs in spiral density waves can be followed self-consistently. Collapse and fragmentation of the clouds will require AMR to resolve scales down to at least 0.001 pc, starting from the galactic disk on a scale of 1 kiloparsec, a dynamic range of 10^6. Up to 20 levels of nested grids might be needed for this resolution. The greatest challenge to such a calculation will be load balancing the AMR algorithms across petascale platforms consisting of millions of processors. Much of the computational effort will be associated with the broader range of physics included in the models and the attendant increase in simulation complexity of the model, as opposed to managing the AMR grids and the data exchanges through the hierarchy of grids. Substantial effort will be required to develop scalable algorithms for self-gravity, radiation transport, cosmic-ray transport, and solvers for the stiff ordinary differential equations (ODEs) associated with the chemistry and microphysics.

Planet Formation

Understanding the formation and evolution of planetary systems in the gas disks that surround protostars is a problem related to understanding star formation. Current theory suggests that planets are formed by one of two routes: either gravitational fragmentation if the disk is massive and cold enough, or through coagulation of solid dust particles into meter-sized "planetesimals," followed by collisional aggregation of planetesimals into protoplanets. The largest protoplanets can capture large amounts of gas from the disk and grow into Jupiter-like gas giant planets. Both processes are very poorly understood. Moreover, in both alternatives the dynamics of the gas disk play a crucial role. It is thought that weakly ionized disks around protostars have complex dynamics and structure. Turbulence can be driven by magnetohydrodynamical instabilities in regions where the gas is sufficiently ionized (most likely the surface layers). Large-amplitude spiral waves and gaps can be cleared by the interaction with growing planets embedded in the disk. Finally, the central protostar can affect the temperature and thermodynamics of the disk via outflowing matter in winds and irradiation by photons. To date, only the most basic theoretical questions have been addressed. With new observational data from Spitzer and from the Atacama Large Millimeter Array (ALMA), a new radio telescope being constructed in Chile with funding from the National Science Foundation (NSF), and with new planet-finding missions such as NASA's Kepler and Terrestrial Planet Finder (TPF) missions, the need to understand planet formation has never been more pressing.

Computations to address these questions will probably consist of three-dimensional magnetohydrodynamic simulations on very large grids (at least 1024^3) that capture the entire disk. Since protostellar disks are very weakly ionized, additional microphysics must be included to capture nonideal magnetohydrodynamic effects. Finally, an understanding of self-gravity will be needed to follow the fragmentation of the disk and of radiation transport to model the thermodynamics realistically. The requisite mathematical models and numerical algorithms are similar to those needed for the related problem of star formation: magnetohydrodynamics, self-gravity, and radiation transport on very large grids, probably with AMR.

The Sun and Stellar Evolution

How do stars evolve? Stellar evolution is a mature field that is now encountering a stiff set of challenges from rapid improvement in astronomical data. Understanding of stellar evolution is also one of the foundations for inferring the evolution of star clusters and galaxies, through age estimates and element production by nucleosynthesis. One-dimensional simulations have been exploited to the limit; three-dimensional simulations including rotation and magnetic fields are beginning to be feasible, thanks to the steady increases in computing power.

What is the true solar composition? Understanding the Sun is intrinsically interesting, but the Sun is also the best-observed star and therefore the most precise test of stellar evolution calculations. Helioseismological observations use the character of sound waves resonant in the solar volume to constrain the run of sound speed and density through deeper layers, almost to the center; the level of precision is better than 1 percent through much of the Sun and far better than for any other star. The internal rotation field of the Sun has been partially traced, and the behavior of magnetic fields in the solar convection zone is being probed. Present theory does not stand up to these challenges: The standard solar model (SSM) of Bahcall and collaborators, the helioseismological inferences for sound speed and density structure, and the best three-dimensional simulations of the stellar atmosphere and photosphere do not agree. One suggestion is that the measured solar abundances of neon and argon are significantly in error. The abundances from the new three-dimensional atmospheres have been challenged. The SSM is a spherically symmetric and static model of the Sun and may itself be the weak link. Three-dimensional simulations of the solar convection zone produce features not represented in the SSM (such as entrainment, g-mode waves, mixing, etc.). Are dynamic effects significant in the evolution of ordinary stars like the Sun? Significant extensions of three-dimensional simulations, which include rotation and magnetic fields, are needed to answer this question, which may be important for understanding both solar physics and stellar evolution.

The First Stars

What followed the Big Bang? After expansion and cooling, the Universe began to form stars and galaxies and in the process yielded the first elements beyond hydrogen and helium. What were these first stars like? Did they produce gamma-ray bursts? Did the first stars make the elemental pattern we observe today in the oldest stars? To simulate them accurately requires a predictive theory of stellar evolution. The elemental yields from such stars are especially sensitive to mixing because there are almost no elements but H and He, so that any newly synthesized element is important. Three-dimensional simulations of turbulence, rotation, binary interaction, and convection, combined with a three-dimensional simulation of thermonuclear burning, are not currently feasible, but they are needed.

If astrophysicists can develop an accurate and predictive theory of star formation that agrees with observations in the current epoch, it is likely they will be able to infer the properties of the first stars that formed in the early Universe. Arguably, understanding the formation of the first stars is easier than understanding star formation today, because the initial conditions were much simpler and because the first stars formed in almost pure H and He gas, in which the heating and cooling processes were much simpler and magnetic fields were probably not dynamically important. However, it is unlikely that we will ever be able to observe the formation of the first stars directly (although we may detect their impact on the surrounding medium). To be confident of the theoretical predictions, we must therefore be certain our numerical models are correct. This requires validation of theory using current epoch data.

Major Challenge 5. Understanding Supernovae and Gamma-Ray Bursts and How They Explode

Our understanding of the birth, evolution, and death of stars is the foundation of much of astrophysics. The most violent end points of stellar evolution are supernova explosions, resulting in the complete disruption of the star or in the collapse of the stellar core to form neutron stars or black holes, accompanied by gamma-ray bursts (GRBs). It is believed that these are the processes that created the chemical elements. Supernovae that are not of Type Ia result from the collapse of the core of a massive star. Despite great progress, core collapse supernovae remain mysterious: There is no clear understanding of why they explode, how they reach the point of explosion, or how the observed explosion properties are related to their history.

Type Ia Supernovae

One of the most spectacular insights from astrophysics is that most of the content of the Universe, being either dark matter or dark energy, is invisible and can be detected only by its effect on the curvature of space-time (that is, by its gravitational effects). This insight comes from our understanding of a particular type of exploding star, a Type Ia supernova. These objects are relatively uniform in behavior, so that their apparent brightness is a crude indicator of their distance from Earth. Careful study of nearby events allowed Mark Phillips to discover that the brightest events last longest; this modest effect allowed the mapping between apparent brightness and distance to be significantly refined. Type Ia supernovae are among the brightest observed events, so this distance scale reaches the limits of observation. Data on Type Ia supernovae, coupled with this distance scale, showed the expansion of the Universe to be accelerating and allowed the amount of dark matter and dark energy to be determined. The argument assumes that the nearby supernovae used to derive the Phillips relations have properties identical to the most distant supernovae. Direct examination of the spectral light from the explosions gives results consistent with this assumption. To improve our understanding, we need to know more about what Type Ia supernovae really are. It is currently thought that they are white dwarf stars, of mass close to the Chandrasekhar limit (1.45 solar masses, the maximum mass that can be supported without a continuing energy input), that ignite and explosively burn. The explosion is supposed to be brought on by accretion of mass from a binary companion.

How do the progenitors of Type Ia supernovae evolve to ignition? What are the properties of the star at ignition? What is the exact mechanism of the explosion? To answer these questions requires, in a critical way, a variety of studies that are dependent on HECC. Much of the understanding of Type Ia supernovae is based on computationally intensive simulations of the radiation and spectra. The identification of specific systems as progenitors of Type Ia supernovae depends on our ability to connect the observed properties to the theoretical models of the evolution of the stellar interior by plausible simulations. Fully three-dimensional simulations of turbulent, self-gravitating, thermonuclear-burning plasma are required; simulations with zoning sufficient to resolve the star down to the inertial range of turbulence are not yet feasible, so a combination of large-scale simulations with subgrid modeling and direct numerical simulation of subsections of the star is needed. Laboratory experiments with high-energy-density lasers and Z-pinch devices can be used to test some of the physics of the simulations; extensive computing is also needed to model the experimental results.

Core Collapse in Massive Stars

Much is unknown about core collapse in massive stars. There is no clear understanding of the nature of the explosion mechanism(s), how collapse proceeds, or of the expected fluxes of neutrinos and gravitational waves. However, two recent observations have revealed details that, if incorporated into new three-dimensional computational models, seem likely to increase our understanding of this fundamental process.

The first set of these observations—of supernova 1987A, the brightest supernova observed since the invention of the telescope—showed striking agreement with theoretical models in some respects (the neutrino flux, for example), but its appearance with three independent rings, ejected during the presupernova evolution, was unexpected. These rings indicate obvious rotational symmetry, and they have been the subject of much speculation. What is the effect of rotation on core collapse? Since such supernovae are not spherically symmetric, they are poorly modeled by one-dimensional treatments. Shear flow in stellar plasma generate magnetic fields, so a three-dimensional magnetohydrodynamic simulation is required for realism. These simulations are very demanding computationally, but they have great promise. There is a wealth of observational data (from pulsars, magnetars, X-ray sources, and supernova remnants) with which such simulations could be compared.

The second set of observations was the discovery that at least some GRBs are related to supernova explosions. These explosions are at cosmological distances, and they require energy supplies much larger than are provided by thermonuclear explosions. However, matter degrades gamma-ray energy, so the fact that GRBs are seen at all suggests that both a core collapse (for energy) and a near vacuum (to protect the gamma rays) are involved. A homogeneous mass distribution would degrade the GRBs and result in (only) a supernova. How does this shift come about? Can it be simulated?

Figure 2-4 illustrates some of the surprises in store as three-dimensional simulations of stars become common. The stage is set for the core collapse of a massive star by the thermonuclear burning of oxygen nuclei. The simulations show new, previously ignored phenomena: The burning is wildly fluctuating and turbulent, and it mixes in new fuel from above and ashes from below the burning layer. Including such effects will change the theory of presupernova evolution. The implications go far beyond this important example. How stars evolve is intimately connected with how much mixing occurs in their interiors, so that all the theoretical understanding of stellar evolution and supernovae will be affected. It is finally possible to make progress with direct three-dimensional simulation; further progress will require the inclusion of rotation and magnetohydrodynamics.

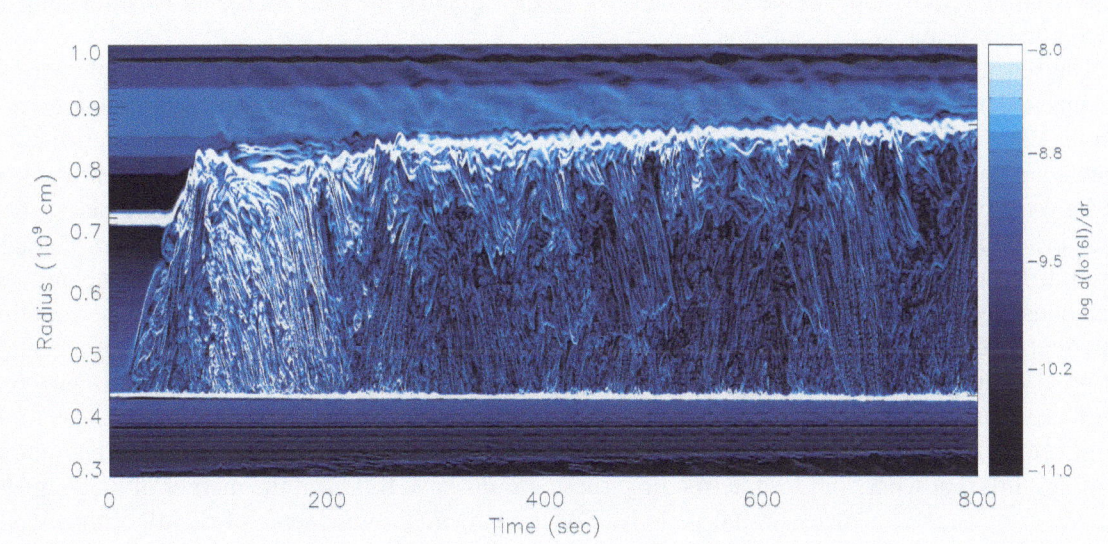

FIGURE 2-4 The evolution of abundance gradients in the convective shell of a 23-solar-mass star shortly before core collapse and supernova explosion. This plot shows the abundance gradient (light colors indicate the highest gradient) as a function of radius and time. This plot was generated from a three-dimensional simulation using a compressible fluid dynamics code that includes multiple nuclear species and their thermonuclear burning. Regions of steep gradient in composition are indicative of turbulent mixing that brings entrained fuel (^{16}O nuclei) into hot regions to be burned. Waves are generated in stably stratified regions above and below the convective shell, and they may be seen beyond the entrainment layers at the convective boundaries. Such oxygen-rich material will be explosively burned by the supernova explosion to form elements from silicon to iron, which will then be ejected into interstellar space, later to be seen in young supernova remnants like Cassiopeia A. Such time-dependent, multiscale, multifluid, three-dimensional simulations are changing the way astrophysicists think about how stars behave. SOURCE: Meakin and Arnett (2007).

Major Challenge 6. Predicting the Spectrum of Gravitational Waves from Merging Black Holes and Neutron Stars

A prediction of Einstein's General Theory of Relativity is that accelerating masses should produce distortions in space-time called gravitational waves. For example, two stars in tight orbit around one another will produce a spectrum of these waves with a frequency determined by their orbital period. Such waves are detectable at Earth as extremely small distortions of space—that is, as a change in the distance between objects as the waves pass by. The United States has invested considerable resources to build the Laser Interferometric Gravitational Observatory (LIGO), an instrument designed to detect gravitational waves from astronomical sources such as close binary stars.

The strongest signals will come from compact objects (such as black holes or neutron stars) in binary systems undergoing mergers. Binary black holes can be formed from the evolution of a binary system containing massive stars. When the stars die they become core-collapse supernovae (see Major Challenge 2), and if their cores are massive enough they will form black holes. The gravitational radiation emitted as the black holes orbit each other removes energy and angular momentum from the system, causing the orbits to decay, so that the stars move ever closer together. The closer they are, the stronger the gravitational radiation and the faster the orbits decay. Finally, the stars merge into a single object in a final burst of gravitational radiation.

The sensitivity required to detect this burst of radiation from merging black holes is staggering; it amounts to detecting changes in the distance between two points several kilometers apart that are smaller than the diameter of a neutron. The sensitivity of LIGO can be greatly increased if the spectrum of the expected radiation is known, so that the observed signal can be correlated with the expected waveform to test for a match. This requires computing the gravitational radiation signal from merging black holes.

In principle, computing this waveform simply requires solving Einstein's equations, a set of coupled partial differential equations that describe the evolution of space-time. However, physicists and mathematicians have struggled with this task for more than 30 years. Given the complexity of the equations, numerical methods are the only possible means of solving the problem. But developing accurate, stable, and reliable numerical algorithms to solve Einstein's equations in three dimensions as a binary black hole merges has proved to be a considerable challenge.

Recently, there was an enormous breakthrough in the field: Several groups have described numerical algorithms that work, and for the first time these groups have followed the merger of black holes in all three dimensions for many orbital periods. The fundamental components of the algorithm include solution of the partial differential equations (PDEs) using simple centered differencing, AMR to capture a range of scales near the merging black holes, constraint damping to limit the accumulation of truncation error, excision of singularities using internal boundary conditions over patches of the grid inside the event horizon, and adoption of a coordinate system that asymptotically extends to infinity to minimize reflection of outgoing waves from the boundaries. Current simulations typically use a grid of 256^3 cells with 10-20 levels of refinement, running on distributed-memory parallel machines with up to several hundred processors. The first gravitational waveforms produced by these mergers have now been reported. Such results are precisely what LIGO and future gravitational wave observatories need to maximize their sensitivity.

With this breakthrough, a whole new field of research is opening up in numerical relativity. First, a large parameter space of black hole mergers must be computed to understand how the outcome is affected when the mass and spin of the initial objects are varied. Next, both matter and electromagnetic fields must be incorporated into the numerical methods so that mergers of neutron stars can be followed. This will require substantial advances in the numerical methods because, although the mathematical properties of Einstein's equations and the equations of magnetohydrodynamics have some important similarities, they are nonetheless quite different. (For example, the equations of magnetohydrodynamics admit discontinuities as solutions, whereas Einstein's equations do not. Accurate and stable numerical algorithms for shock capturing are therefore required for magnetohydrodynamics.) Finally, as the simulation tools mature, they can provide the foundation for an enormous effort to extract a physical understanding of high-energy phenomena where relativistic effects are important (which goes well beyond the merger of binary systems to include core-collapse supernovae).

Physics is at the beginning of a renaissance in the study of general relativity. The first generation of numerical tools required to solve Einstein's equations are now available. Much work remains to be done, and all of it relies on computation.

METHODS AND ALGORITHMS IN ASTROPHYSICS

A wide variety of mathematical models, numerical algorithms, and computer codes are used to address the compelling problems in astrophysics. This section discusses some of the most important, organized by the mathematics.

- *N-body codes*. Required to investigate the dynamics of collisionless dark matter, or to study stellar or planetary dynamics. The mathematical model is a set of first-order ODEs for each particle, with acceleration computed from the gravitational interaction of each particle with all the others. Integrating particle orbits requires standard methods for ODEs, with variable time stepping for close encounters. For the gravitational acceleration (the major computational challenge), direct summation, tree algorithms, and grid-based methods are all used to compute the gravitational potential from Poisson's equations.
- *PIC codes*. Required to study the dynamics of weakly collisional, dilute plasmas. The mathematical model consists of the relativistic equations of motion for particles, plus Maxwell's equations for the electric and magnetic fields they induce (a set of coupled first-order PDEs). Standard techniques are based on particle-in-cell (PIC) algorithms, in which Maxwell's equations are solved on a grid using finite-difference methods and the particle motion is calculated by standard ODE integrators.
- *Fluid dynamics*. Required for strongly collisional plasmas. The mathematical model comprises the standard equations of compressible fluid dynamics (the Euler equations, a set of hyperbolic PDEs), supplemented by Poisson's equation for self-gravity (an elliptic PDE), Maxwell's equation for magnetic fields (an additional set of hyperbolic PDEs), and the radiative transfer equation for photon or neutrino transport (a high-dimensional parabolic PDE). A wide variety of algorithms for fluid dynamics are used, including finite-difference, finite-volume, and operator-splitting methods on orthogonal grids, as well as particle methods that are unique to astrophysics—for example, SPH. To improve resolution across a broad range of length scales, grid-based methods often rely on static and adaptive mesh refinement (AMR). The AMR methods greatly increase the complexity of the algorithm, reduce the scalability, and complicate effective load-balancing yet are absolutely essential for some problems.
- *Transport problems*. Required to calculate the effect of transport of energy and momentum by photons or neutrinos in a plasma. The mathematical model is a parabolic PDE in seven dimensions. Both grid-based (characteristic) and particle-based (Monte Carlo) methods are used. The high dimensionality of the problem makes first-principles calculations difficult, and so simplifying assumptions (for example, frequency-independent transport, or the diffusion approximation) are usually required.
- *Microphysics*. Necessary to incorporate nuclear reactions, chemistry, and ionization/recombination reactions into fluid and plasma simulations. The mathematical model is a set of coupled nonlinear, stiff ODEs (or algebraic equations if steady-state abundances are assumed) representing the reaction network. Implicit methods are generally required if the ODEs are solved. Implicit finite-difference methods for integrating realistic networks with dozens of constituent species are extremely costly.

For all of these methods the main computational challenges are the enormous number of particles (10^{10-11} at present) and the large grids (as big as 2048^3, or 10^{10} cells, at the moment, with larger calculations planned for the future). Complex methods can require 10^3 or 10^4 flops per cell per time step and generate hundreds of gigabytes of data in a single snapshot. Floating point performance is often limited by cache access and the speed of the on-chip bus. Some algorithms (for example, grid-based fluid dynamics) scale very well to tens of thousands of processors, while others (for example, elliptic PDEs, such as Poisson's equation) require global communication, which can limit scaling.

HECC FOR DATA ANALYSIS

As indicated earlier, four of the six questions identified in this chapter as the major challenges facing astrophysics in the coming decade require HECC to test theoretical models against observational data. Without HECC, there might be little or no progress in these areas of astrophysics.

For the other two major challenges identified earlier (the nature of dark matter and dark energy), the most productive mode of investigation will likely be observation, which will collect massive amounts of data. For example, the largest survey of the sky to date, the Sloan Digital Sky Survey, has generated about 2.4 TB of data over the past 5 years. Over the next 5 years, the PanSTARRs survey will generate 20-200 TB of data, with an image archive that may grow to 1.5 petabytes (PB). The Large Synoptic Survey Telescope, an 8-meter-class telescope that will provide the deepest survey of the sky, will generate 30 PB of data over the next 10 years. In addition, the scale of numerical computation envisioned in the future will also generate massive data sets. Current simulations of galaxy formation already generate 30 TB of data per run. This will grow to 1 PB per simulation in the near future.

Management, analysis, mining, and knowledge discovery from data sets of this scale are a challenging problem in HECC in their own right. For data-intensive fields like astronomy and astrophysics, the potential impact of HECC is felt not just in the power it can provide for simulations but also in the capabilities it provides for managing and making sense of the data. Because the amount, complexity, and rate of generation of scientific data are all increasing exponentially, even some research challenges that are primarily addressed through observation are also critically dependent on HECC.

Systems that deal with data on this scale at present usually need to access thousands, perhaps even hundreds of thousands, of disks in parallel; otherwise the system's performance would be limited by the rate of input/output (I/O). Thus scalable parallel I/O software is extremely critical in such situations. There are many specific challenges within this domain and barriers to progress, including the following:

- Are there data management tools available that can manage data at the petascale while allowing access by scientists all over the world? For example, data obtained from sky surveys currently range from 1 to 2 PB a year, with the amount likely to increase as more sophisticated instruments with much higher resolutions are deployed. Similarly, as petascale simulations become feasible, many astrophysics calculations will produce multiple petabytes of data.
- What data management models are good on these scales? A traditional model assumes that the data are moved (for example, as the result of a query) to the place of their end use, but that model does not seem to be scalable. New models and architectures are needed that enable the querying and analysis of data to be distributed to the systems that store and manage data.
- Data volumes produced by simulations or observations are rising very fast, while computational and analytical resources and tools that are needed to perform the analysis are lagging far behind in terms of performance and availability. Ever-more-sophisticated algorithms and computational models result in larger data sets, but the bandwidth available to deal with these data is not keeping up. Thus, parallel and scalable techniques for analysis and handling data are needed. Online parallel algorithms that can guide the analysis and steer simulations have the potential to help. Scalable parallel file systems and middleware that can effectively aid the scientists with these massive volumes of data are critical.
- Traditional analysis techniques such as visualization are not scalable, nor are they suitable for knowledge discovery at petascale ranges, although they may enable the user to guide the knowledge-discovery process. Are there statistical and data-mining tools that can help scientists to make

discoveries from massive amounts of data? Can these tools be scaled appropriately? Data-mining methods such as clustering, neural networks, classification trees, association rules discovery, and Bayesian networks allow scientists to automatically extract useful and actionable patterns, representing information and knowledge, from data sets. They need to be adapted or developed for analyzing scientific applications so that they can enhance the scientists' ability to analyze and learn from their simulations and experiments.

- The size of simulations in astrophysics, as described earlier, is reaching the petascale regime. Storing and retrieving the data and results for subsequent analyses are as big a challenge as is performing the computations. The challenge is to extract and share knowledge from the data without being overwhelmed by the task of managing them. How can scientists interact with their data under these conditions?
- Traditionally, scientists have done simulations or experiments and then analyzed them and published the results, which are subsequently used by others. An emerging model relies more on teams working together to organize new data, develop derived data, and produce analyses based on the data, all of which can be shared with other scientists. Thus more and larger scientific databases are being created. What are the exemplars for such sharing, and how can other scientists use these data sets to accelerate their knowledge discovery? Ontologies and taxonomies must be developed to enable that knowledge discovery.
- There are no guidelines for building or configuring balanced systems in terms of computation and I/O requirements. There are no benchmarks (analogous to those used to measure, say, the "top 500" most-powerful supercomputers) for balanced systems that incorporate I/O, storage, and analysis requirements. Would it be possible to build such benchmarks for balanced systems?

REALIZING THE POTENTIAL IMPACT OF HECC ON ASTROPHYSICS

Astrophysics is a computationally mature discipline with a long history of using computing to solve problems. To a large extent, the astrophysics community writes its own codes, and many astrophysicists are knowledgeable about programming issues on modern HECC architectures. Most computation in astrophysics uses a mix of numerical methods, including grid-based methods for hydrodynamics and magnetohydrodynamics, AMR schemes, particle-based methods for N-body and plasma dynamics, and methods for radiation transport using both grids and Monte Carlo algorithms. There are not enough researchers in the community to support a true community-code model, as is the case in atmospheric sciences, for example. Instead, individual groups of a few researchers develop alternative algorithms, and comparing the results of different groups is very beneficial to the research. Students in astrophysics are trained in computation at both the undergraduate and the graduate levels, and researchers trained in computation are heavily recruited. Most large-scale research computing in astrophysics is performed at the NSF-funded supercomputing centers or in university-owned (medium-sized) facilities. A few individual departments are fortunate enough to provide small- to medium-sized facilities to their members. Since such a large fraction of astrophysical research uses HECC, there seems never to be enough resources available, and astrophysics would certainly benefit from more access to much larger facilities. Such access should come with strong support of code and algorithm development efforts to add new physics and exploit new architectures. For all these reasons, progress in astrophysics is intimately tied to progress in HECC.

On the hardware side, the components of HECC infrastructure that are most necessary for progress in astrophysics are (1) an increase in processing capabilities (flops), whether through faster processors or, for many computations, increased parallelism, (2) better I/O performance that is matched to the floating point performance, and (3) hardware that can handle massive data sets. It is worthwhile to note that the hardware is in some ways the easiest need to satisfy.

On the software side, the progress-limiting steps for astrophysics are (1) better methods for non-ideal magnetohydrodynamics, radiation transfer, and relativistic gas dynamics with general relativity (for a few examples) and (2) better visualization tools and better optimization and profiling tools that can make the codes run faster.

Certain common components, such as the interface to AMR tools, file formats, and visualization tools, would benefit from common community standards. For many reasons—for example, the wide variety of physics involved in astrophysical research (from N-body dynamics to magnetohydrodynamics); the small size of the astrophysics community in comparison with other communities such as atmospheric sciences; and the evolving nature of the mathematical models used to describe astrophysical systems (from hydrodynamics to magnetohydrodynamics and, finally, weakly collisional plasma dynamics)—the development of true community codes is not likely in astrophysics. Nonetheless, the need for community *standards* is pressing.

The astrophysics research community needs support for porting its own codes to petascale environments. No one knows which of the current algorithms and software will scale, because only a few astrophysics codes have scaled beyond a few hundred processors. Even when the mathematical models are known, transitioning them to a larger number of processors or new processor architectures presents challenges, some of them very difficult.

Finally, it is important to emphasize that the most critical resource in the research enterprise, and the one that is always rate limiting, is the number of highly qualified personnel trained in HECC. Better training means the community can make more optimal use of existing resources. Thus, education in computational science must be emphasized at every level. There needs to be support for people whose expertise is at the boundary of computer science and the application discipline. People who really know how to make use of HECC capabilities are scarce and essential.

To conclude, the committee identified the following likely ramifications of inadequate or delayed support of HECC for astrophysics:

- The rate of discovery would be limited. Those foregone discoveries would probably have enriched our fundamental understanding of the Universe and could have provided tangible benefits: For example, understanding how the atmospheres of Venus and Mars and of Titan (the largest moon of Saturn) affect the global climates of those bodies would be expected to help us model the changing climate of Earth.
- Inadequate support for HECC would lead to a shortage of training for highly qualified personnel.
- Inadequate support for HECC would limit our ability to capitalize on the investments in expensive facilities. Major observational facilities, especially those in space, can cost billions of dollars. Guidance from theoretical and computational modeling on how best to observe systems can dramatically increase the success rate of such facilities. For example, the LIGO requires templates of expected gravitational waveforms to distinguish signals from the noise. An accurate library of waveforms, which can only come from computation, could mean the difference between whether LIGO does or does not detect a signal.
- Some data are likely to be underexploited. Without enough HECC, much of the data from big surveys is unlikely to be processed in any meaningful way, which means losing information from which new discoveries might come. For example, detecting near-Earth asteroids requires high-time-resolution surveys over a large area of the sky. Although the likelihood of an asteroid

impacting Earth in the foreseeable future is extremely small, it does not seem wise to reduce our chance of detecting such a threat by failing to process that survey data.

REFERENCES

Bate, M.R., and I.A. Bonnell. 2004. Computer simulations of star cluster formation via turbulent fragmentation. In *The Formation and Evolution of Massive Young Star Clusters*. H.J.G.L.M. Lamers, L.J. Smith, and A. Nota., eds., ASP Conference Series, 322. San Francisco, Calif.: Astronomical Society of the Pacific, p. 289.

Faucher-Giguere, C.-A., A. Lidz, and L. Hernquist. 2008. Numerical simulations unravel the cosmic web. *Science* 319:52.

Heitsch, F., M.-M. Mac Low, and R.S. Klessen. 2001. Gravitational collapse in turbulent molecular clouds II: MHD turbulence. *The Astrophysical Journal* 547: 280.

Krumholz, M.R., R.I. Klein, and C.F. McKee. 2007. Radiation-hydrodynamic simulations of collapse and fragmentation in massive protostellar cores. *The Astrophysical Journal* 656: 959.

Meakin, Casey A., and David Arnett. 2007. Turbulent convection in stellar interiors, I: Hydrodynamic simulation. *The Astrophysical Journal* 667 (1): 448-475.

NRC (National Research Council). 2001. *Astronomy and Astrophysics in the New Millennium*. Washington, D.C.: National Academy Press.

NRC. 2003. *Connecting Quarks with the Cosmos: Eleven Science Questions for the New Century*. Washington, D.C.: The National Academies Press.

NSTC (National Science and Technology Council). 2004. *The Physics of the Universe: A Strategic Plan for Federal Research at the Intersection of Physics and Astronomy*. Washington, D.C.: NSTC.

Springel, Volker, Simon D.M. White, Adrian Jenkins, et al. 2005. Simulations of the formation, evolution and clustering of galaxies and quasars, *Nature* 435: 629.

3

The Potential Impact of HECC in the Atmospheric Sciences

INTRODUCTION

Weather and climate events, and now long-term environmental change, are increasingly significant in both private and public decision making. We require warnings of severe and damaging weather, predictions of the tracks and intensity of hurricanes and winter storms, and longer-term forecasts about seasonal variations. Extreme event theory has been extended into a statistical characterization of droughts, floods, heat waves, and other environmental forces. In recent years we have sought to foresee the global changes that might be in progress, anthropogenic or otherwise.

The skill and reliability of forecasts has increased markedly since the advent of weather radar, Earth-observing satellites, and powerful computers. Today tornados are almost always seen on radar in time to issue warnings, and satellite images portray the growth and trajectory of hurricanes and other storms. To look beyond a few hours, we combine powerful computers with the relevant laws of physics converted into mathematical models to predict how the observed present state of the global atmosphere will evolve in the hours, days, months, and years ahead.

The significance of the decisions and planning that depend on weather and climate forecasts and simulations justifies efforts to make dramatic improvement. We expect that this improvement will be achieved as the numerical models that portray events in the atmosphere and ocean and on the land surface gain significantly in their resolution and sophistication. Detailed calculation of the feedbacks in the climate system, including the carbon cycle and other elemental cycles, are required for accurate climate prediction. Numerical weather forecasting in particular is approaching the ability to take account of local features (for example, lakes, ridges) and local variations in atmospheric moisture content, and the successful modeling of atmospheric variables on these local scales should lead to a significant improvement in forecast quality.

The importance to society of weather and climate information is evident from the significant investments in surface and balloon observations and in radar and satellite observing systems. The observations themselves provide insight into present conditions and expectations for the next few hours, but the larger value of those observations is realized with the computer models that turn them into longer-range forecasts. This forecast process depends on scientific understanding of complex physical and chemical processes in the atmosphere and detailed simulation of momentum, heat, energy, and molecular exchange between the land, ocean, and atmosphere. The radiant energy from the Sun is converted into thermal energy and then into the kinetic energy of winds and storms, all involving a highly nonlinear exchange of energy among small- and large-scale phenomena and a continuing exchange of water and chemical constituents between the atmosphere, oceans, and land surface. Because of these continuous exchanges, it is difficult to talk about atmospheric science without also considering ocean, land, and ice.

Skill in atmospheric prediction builds on scientific understanding converted into the mathematical expressions that become a numerical model of Earth and its atmosphere, oceans, and land surface, including detailed treatments of land and ocean biogeochemistry as well as the radiative impacts of atmospheric aerosols and gas-phase chemistry. The ocean is the planet's reservoir of thermal energy and water, the land surface alters energy fluxes, and plants in both the ocean and on land maintain the oxygen balance. The atmosphere is the high-speed transport system that links them together and drives toward a thermodynamic equilibrium that is never attained. The central task facing atmospheric scientists is to unite sufficiently powerful science and sufficiently powerful computers to create a numerical counterpart of Earth and its atmosphere with the similitude required to manage weather, climate, and environmental risk with confidence in the years ahead.

This chapter identifies the major frontier challenges that the atmospheric sciences are attacking in order to realize that central task. In order to identify those challenges, the committee relied on several recent reports, including the following:

- European Centre for Medium-Range Weather Forecasts, 2006, *ECMWF Strategy 2006-2015.* Available at http://www.ecmwf.int/about/programmatic/ 2006/index.html.
- Intergovernmental Panel on Climate Change (IPCC), 2007, *Climate Change 2007: The Physical Science Basis,* Working Group I report, Cambridge University Press.
- National Research Council (NRC), 1998, *The Atmospheric Sciences Entering the Twenty-first Century*, Washington, D.C: National Academy Press.
- NRC, 2000, *From Research to Operations in Weather Satellites and Numerical Weather Prediction: Crossing the Valley of Death*, Washington, D.C.: National Academy Press.
- NRC, 2001, *Effectiveness of U.S. Climate Modeling,* Washington, D.C.: National Academy Press.
- NRC, 2002, *Abrupt Climate Change: Inevitable Surprises*. Washington, D.C.: National Academy Press.
- National Science Foundation (NSF), 2000, *NSF Geosciences Beyond 2000: Understanding and Predicting Earth's Environment and Habitability.* Available at http://www.nsf.gov/geo/adgeo/geo2000.jsp.
- University Corporation for Atmospheric Research (UCAR). 2005. *Establishing a Petascale Collaboratory for the Geosciences: Scientific Frontiers* and *Establishing a Petascale Collaboratory for the Geosciences: Technical and Budgetary Prospectus.* Technical Working Group and Ad Hoc Committee for a Petascale Collaboratory for the Geosciences. Available at http://www.geo-prose.com/projects/petascale_tech.html.

Members of the committee also consulted the peer-reviewed literature in the atmospheric sciences, especially the references listed at the end of this chapter. The committee had extensive discussion with invited guests in a daylong workshop, details of which are included in Appendix B. Further discussions were undertaken with the directors and senior executives of the National Centers for Environmental

Prediction (NCEP) of the National Weather Service and the Geophysical Fluid Dynamics Laboratory (GFDL), both of which are part of the National Oceanic and Atmospheric Administration, and with the director of the European Centre for Medium-Range Weather Forecasts (ECMWF). Particular assistance was received from Louis Uccellini, Ben Kyger, and Steve Lord from NCEP, Brian Gross from GFDL, Dominique Marbouty of the ECMWF, James Hack from the National Center for Atmospheric Research (NCAR), and other colleagues. Special thanks go to Jeremy D. Ross, Storm Exchange, Inc., for assistance with the discussion of challenges and techniques of high-resolution mesoscale modeling.

MAJOR CHALLENGES IN THE ATMOSPHERIC SCIENCES

This section identifies the major challenges facing the atmospheric sciences. Each challenge is given a ranking of [1], [2], or [3], representing the committee's consensus on the degree to which advances in HECC will impact progress. A ranking of [1] indicates that progress would immediately accelerate if advances in HECC were available. Other resources might be partially limiting, too, such as field or laboratory instruments, personnel, or funds for field programs. A ranking of [2] indicates that HECC is currently playing or will soon play a key role, but that other factors, such as immaturity of the models or data sets, are more limiting because they prevent advances in HECC from having an immediate impact. A ranking of [3] indicates that HECC will probably not be a limiting resource within 5 years because the models, data sets, and theories are not fully mature. Even for the challenges ranked [3], however, some computer-intensive modeling or data analysis activities are already under way.

In the brief discussion that accompanies each challenge, references are made to the physical or mathematical aspects of the problem that necessitate attacking it with HECC. These aspects are described in more detail in the section on computational challenges in the atmospheric sciences.

Major Challenge 1: Extend the Range, Accuracy, and Utility of Weather Prediction [1]

Many sectors of the economy and the public at large have come to depend on accurate forecasting of day-to-day weather. Examples include agriculture, transportation, energy, construction, and recreation. Over the last 50 years, the accuracy and range of weather forecasting have steadily improved, owing in equal measure to underlying improvements in

- Physical models of the atmosphere;
- Algorithms for solving partial differential initial- and boundary-value problems;
- Coverage and quality of data for model initialization;
- Algorithms for model initialization; and
- Computers that are faster and have larger memories, allowing higher spatial resolution and running of ensembles of models.

These advances are the result of massive, sustained research efforts in atmospheric physics and dynamics, instrumentation, the mathematics of chaos, and several areas of data filtering and numerical analysis, all of them capitalizing on concurrent progress in supercomputers themselves. While today's weather forecasts are good by previous standards, several serious limitations can nonetheless be noted:

- Even a short-term forecast for one or two days out may occasionally miss a significant weather event or misjudge its strength or trajectory.

- Beyond five days, the reliability (and therefore the utility) of weather forecasts decreases rapidly.
- Quantitative precipitation forecasts are not as skillful as forecasts for other aspects of weather—wind, pressure, and temperature.
- Certain types of severe weather—namely, those triggered by instabilities—are poorly predicted.
- Most weather forecasts do not provide truly local information—down to the scale of a few kilometers—which is the level of specificity needed by some human activities.

While some forecast improvement could be achieved fairly readily just by upgrading the computing platforms, addressing the limitations just listed would require progress on multiple fronts, such as theory, modeling of small-scale phenomena (especially moisture processes), observations, data assimilation and exploitation, and understanding of atmospheric chemistry. The role that HECC is expected to play in this progress is described in the next section.

Major Challenge 2: Improve Understanding and Timely Prediction of Severe Weather, Pollution, and Climate Events [1]

The need for accurate weather forecasts becomes especially acute when natural hazards threaten. At least three types of natural hazard arise in the atmosphere:

- Severe storms (tornadoes, floods, hurricanes, snowstorms, and ice storms);
- Severe climate events (drought, heat wave); and
- Atmospheric response to external events (fire, volcano, pollution release).

The committee decided to separate these extraordinary forecasting challenges from the day-to-day variety described in Major Challenge 1 because severe events involve different physical processes or require a special type of forecast model or a special kind of input data. The time criticality of the latter kind of forecast may also differ. In the case of tornados and flash floods, gains of a few minutes in warnings may make a substantial difference in the value of the forecast, giving the public more time to seek shelter. With hurricanes and ice storms, timely predictions can alert government agencies and power companies to mobilize their emergency crews, people can be evacuated, and ships at sea can alter their routes or prepare for severe conditions. The accurate prediction of extreme events often requires input from local observing systems that may not be part of the standard national or global data system. It is now conceivable—though not yet achievable—that minute-by-minute Doppler radar data could be used to initialize or refine a high-resolution, three-dimensional fluid dynamics model (grid cell of 10 meters) of a developing severe tornadic thunderstorm. Such simulations would require that data ingestion and its forward integration be done substantially faster than in real time, which cannot be accomplished with current high-end computing capabilities.

Droughts and heat waves last longer, several weeks, and they too demand special forecasting methods. Inputs of information on soil moisture and the state of vegetation may be critical, and models with more detail about feedback between the atmosphere and land surface are required. One activity under way in the climate modeling community is an evaluation of the length and magnitude of droughts in climate models over the next 100 years. Initial indications are that droughts will be more intense and last longer in the future and that the inclusion of accurate detailed information on carbon and water exchange between the land, oceans, and atmosphere is critical to the accuracy of the climate simulations.

When dangerous material is unexpectedly released into the atmosphere, it is essential to quickly predict the spread and concentration of the substance. Local turbulence models are required, coupled with current or forecast winds and clouds. In the case of forest fires, the development of the fire itself can be modeled with the appropriate equations, including those for the combustion properties of the fuel.

What is commonly needed for addressing Major Challenge 2 are computational models with finer resolution that incorporate, in most of the cases, models of additional physical interactions. Progress against that common need requires advances in HECC capabilities.

Major Challenge 3: Improve Understanding and Prediction of Seasonal-, Decadal-, and Century-Scale Climate Variation on Global, Regional, and Local Scales [1]

While detailed weather forecasting cannot be extended beyond a month, climate may usefully be forecasted for several months, even into the next season. This is possible because Earth's climate system has some slow components that are less chaotic. The slow components nudge the mean properties of the fast chaotic weather systems. Examples include heat storage in the ocean, snow cover, and albedo changes due to seasonal vegetation variations. Most of the large national and international numerical prediction centers run ensemble models linking atmospheric, oceanic, and land components to provide outlooks for seasonal anomalies up to nine months in advance.

Special attention is paid to El Niño events, in which the sea surface temperatures of the tropical eastern Pacific rise by a few degrees, or La Niña events, in which they cool, modifying the mean state of the climate in a large region, including much of North America. An El Niño prediction thus carries information about the probability of wetter or drier conditions that can be used by agricultural interests and water resource planners in the United States. The science of El Niño prediction is still developing. The El Niño forecasts produced by dynamical models and by statistical means at a variety of institutions show considerable scatter. Nevertheless, the main dynamical models are fairly consistent in identifying the onset of both El Niño and La Niña events with lead times of a few months. They are considerably less successful in anticipating the intensity and duration of those events, however.

The computational challenge related to seasonal forecasts, including El Niño and La Niña events, is clear. A good model must include velocity and temperature fields in the atmosphere and ocean, as well as the coupling between the atmosphere and the ocean and land by turbulent transport. Cloud and radiation processes in the atmosphere are also involved. Ensemble methods are required to address the chaotic nature of the coupled atmosphere-ocean system. Progress will require the pursuit of various techniques by competing research groups. HECC plays a key role in both research and operational prediction, but progress on basic physics is also required.

Major Challenge 4: Understand the Physics and Dynamics of Clouds, Aerosols, and Precipitation [2]

While many aspects of the atmosphere are complex, many scientists agree that clouds and precipitation, and their interaction with aerosols, pose the greatest difficulty. Clouds have an impact on all scales (Stephens, 2005): They can lead to a local thunderstorm that rains on a picnic, or they can generate deadly tornados and hail. The degree of cloud coverage worldwide controls the average Earth albedo, which influences global temperature.

The difficulty in understanding and predicting cloud processes is a result of the multiplicity of aerosol particles and condensed water particles that interact inside clouds as well as the tendency for clouds to encase vigorous turbulent motions. Summer convective clouds often form in response to a sensitive instability in which small thermals of rising air accelerate upward if sufficient latent heat is released from condensing water vapor. In present cloud models, all liquid and ice particles are binned into a small number of categories (between 2 and 10, typically). The transference of material between these categories is represented by crude semiempirical equations, so-called parameterizations. Detailed atmospheric chemistry simulations are required to portray aerosol and radiation impacts, and these in turn require advances in modeling and in HECC.

Even so, major breakthroughs in modeling have occurred. The first numerical simulations of a tornadic thunderstorm were accomplished in the late 1980s. Recently, some success in predicting the amount of cloud cover in stratocumulus decks over the oceans was reported. While these are not per se successful forecasts, they show that the understanding of clouds is advancing.

At the present time, however, forecasts of summer precipitation are done on a probabilistic basis rather than a yes-no basis. Statistically, the reliability of summer precipitation forecasts is much poorer than for other weather forecasts. Even in a simpler situation, such as precipitation from orographic clouds in mountainous terrains, existing models may vary in their prediction of precipitation by factors of two or more.

Progress in understanding cloud physics will require a mix of theory, laboratory and field measurement, and trials with high-resolution numerical models. And those numerical models must include a better representation of the spectrum of particle sizes and more realistic transfer processes, which will multiply the computational difficulty.

Major Challenge 5: Understand Atmospheric Forcing and Feedbacks Associated with Moisture and Chemical Exchange at Earth's Surface [1]

While early theories and numerical models of Earth's atmosphere used a static representation for the lower boundary ("Earth's surface"), the degree of active interaction between ground and atmosphere is now widely appreciated. These surface feedbacks take several forms. On a small scale, a patch of rain will moisten the soil and increase the evaporation, reducing the temperature and altering the winds. On a larger scale, a local drought may kill the vegetation and alter the local roughness and albedo. On a global scale, increased temperature may alter the soil chemistry and release stored soil carbon to the atmosphere. Reduction in sea ice will reduce the albedo and allow massive heat fluxes to emerge from the ocean into the atmosphere. Surface feedbacks have been identified as components of thunderstorm initiation, heat waves and droughts, dust storms, Arctic ocean warming, and other phenomena.

For climate modeling, two important human inputs—fossil fuel emissions and land cover change—are thought to contribute significantly (~8-10 petagrams (Pg) of carbon per year) to atmospheric CO_2. Meanwhile, the world's oceans are a carbon sink (~2 Pg/yr), and another 2 Pg/yr of carbon is lost to a hypothesized "missing sink." Feedbacks also exist between warming and greenhouse gas production. New climate simulations are beginning to quantify the effects of land cover and fossil fuel emissions on the global carbon and water cycles and to identify and quantify the feedbacks among the driving biogeochemical processes and the climate system in the past, present, and future (Cox et al., 2000). Exemplifying today's capabilities, Figure 3-1 shows the result of a simulation in which the physical climate system is fully interacting with the global carbon cycle (Thompson et al., 2004). By incorporating the coupling and feedbacks, the simulation exposes qualitative behaviors that we seek to understand.

To credibly evaluate feedbacks in the climate system, one must couple (1) a physical ocean model that has an embedded biological ecosystem model with (2) an atmospheric physical model embedded with a detailed model of atmospheric particle and gas-phase chemistry. This coupled model must at the same time be further coupled to a detailed model of continental biological and physical processes. The terrestrial biosphere, oceanic biosphere, and classically constructed physical climate system all interact and contribute to the simulated climate. The biogeochemical and physical climate system is fully coupled

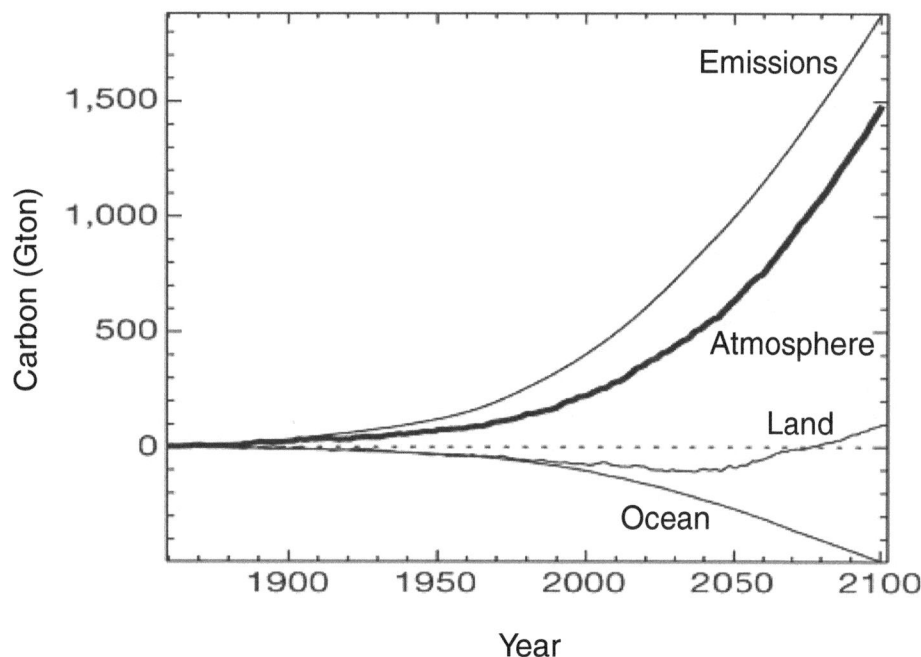

FIGURE 3-1 Output from a coupled carbon-cycle/physical-climate simulation that allows the exchange of anthropogenic CO_2 between atmosphere, ocean, and land. (Derived from the simulations of Thompson et al., 2004.) The vertical axis displays the cumulative gigatons of anthropogenic carbon added to each of the reservoirs shown, with negative numbers representing absorption from the atmosphere. This indicates that the ocean is taking up a fraction of the anthropogenic CO_2 and that the land does this as well until about 2070. The curve labeled "emissions" displays the integrated anthropogenic output. Until about 2070 the land system is a slight sink for CO_2, but carbon feedbacks inside the model lead it to become a source thereafter. This directly impacts the atmospheric concentration of CO_2, but current computing capabilities limit our ability to discern the strength and timing of this predicted inflection in terrestrial uptake. Coupled models such as this are essential for understanding climate, and HECC resources are essential to enable the incorporation of enough geochemical and atmospheric chemistry detail.

in the sense that changes and trends in the biogeochemistry influence the physical climate at each time step and vice versa.

Clearly, this is a daunting computational task. While we know enough of the underlying science in atmospheric chemistry, aerosols, terrestrial biogeochemistry, and ocean biogeochemistry to create fully coupled models, and we have adequate mathematical models for those processes, it is not yet possible to exploit all these components because of limitations of HECC. At present, most fully coupled production simulations cannot even use rudimentary models of all the interacting processes. And so, while Figure 3-1 allows us a glimpse of critical understanding, advances in HECC are needed to better represent the underlying science and improve our understanding.

Major Challenge 6: Develop a Theoretical Understanding of Nonlinearities and Tipping Points in Weather and Climate Systems [2]

From a theoretical viewpoint, a nonlinear system such as Earth's climate system can clearly undergo sudden changes in state, even with slow changes in external forcing. Positive feedback processes such as ice-albedo feedback, water vapor feedback, or warming CO_2 feedback could all push toward a bifurcation or tipping point. Even weather systems such as severe thunderstorms or cyclones might behave like this. A theoretical understanding of nonlinear dynamics is therefore essential.

The matter is made more urgent by evidence that Earth's climate may already have experienced abrupt change in the form of glacial retreat (NRC, 2002). Another possible example of change is the postulated shutdown of thermohaline circulation in the Atlantic Ocean.

A potential danger in studies of nonlinear behavior is that low-order models with few degrees of freedom seem to overpredict the occurrence and magnitude of bifurcations. When more components of a system are included in a model, more competing pathways of change are present and abrupt shifts are less likely. Thus the reliable prediction of bifurcation requires complex, multicomponent models.

Major Challenge 7: Create the Ability to Accurately Predict Global Climate and Carbon-Cycle Response to Forcing Scenarios over the Next 100 Years [1]

As concerns over global warming grow, it is important to accurately predict global climate as a function of various emission scenarios. Two recent reports (IPCC, 2001 and 2007) show that such predictions are affecting public opinion and industrial and government policy. Features of these predictions include the following:

- Regional climate and economic impacts of climate change, including scenarios for adapting land use and agriculture.
- Impacts on ocean circulation and productivity.
- Impacts on snow, water storage, and river flow.
- Inclusion of the many feedbacks connecting the global carbon cycle, water vapor, and biogeochemical cycles.
- Detailed simulations of the time trends in the oxidative state of the atmosphere via comprehensive atmospheric chemistry.
- Interdependency between economic decisions and greenhouse gas emissions.

This major challenge builds on the results of Major Challenge 5. The potential impact of HECC advances on Major Challenge 7 would stem both from improvements to and testing of computer models and from allowing models to be run over tens of simulated decades, which requires immense amounts of supercomputer time. Because our understanding of physical, chemical, and biological processes is adequate, progress would come as soon as the advances in HECC are introduced. In the longer term, the development of computational models that can resolve all-important regional differences will also require significant work on algorithms. That is because the algorithms for various components of the fully coupled simulations scale in different ways, so that the complete simulation will not scale readily to the tens of thousands of processors anticipated with next-generation supercomputers.

Given these needs, the credibility, accuracy, and availability of global climate models must continue to increase. To retain its credibility, the most recent IPCC report (2007) limited its discussion to aspects in which there is broad agreement across models. The models predicted a general warming of between 2 and 6 degrees over the next 50 to 100 years, somewhat accelerated at higher latitudes. Precipitation increases are predicted for the middle and high latitudes, with drying in the subtropics.

The predictions about climate change on the regional scale are less consistent. Models differ considerably on whether regions such as Western Europe, the southwestern United States, or Australia will become wetter or dryer and on how such changes will impact agriculture. The model-dependence of these predictions is probably due to how local surface feedbacks (for example, sea surface temperature, soil moisture, and vegetation) are simulated or to subtle resonances related to the location and amplitude of planetary waves in the upper atmosphere. Regional feedbacks between the biogeochemical-carbon cycle and the regional hydrologic cycle are critical to the computed climate, and models for these contain significant uncertainties. These uncertainties call for fundamental scientific research to improve understanding. Additional computational resources will also be required for data analysis given the massive amounts of data that can be produced by climate models and by the newer generations of satellites that will be deployed in the decades ahead.

The impact of global warming on agriculture, ocean fisheries, and water resources is expected to be significant. Simulations of the next 100 years will require a detailed understanding of how ocean temperature, salinity, and nutrient distributions will evolve over time. The prediction of these impacts requires special models that are still not well developed.

The demands on computing resources for climate studies are huge, particularly because runs are so long. Typically, because weather is so variable, at least 5 years of weather must be simulated to identify moderate shifts in climate. In typical cases, thousands of years of simulated time are required to allow all the different pools of heat and carbon to reach a steady state. In addition, high spatial resolution is required to capture the dominant physical processes. Some climate simulation runs take as much as several months of wall clock time, but longer runs are not practical. Thus the most ambitious climate simulations are unlikely to be undertaken unless HECC resources are available.

Major Challenge 8: Model and Understand the Physics of the Ice Ages, Including Embedded Abrupt Climate Change Events Such as the Younger Dryas, Heinrich, and Dansgaard-Oeschger Events [2]

During the last 2 million years (the Pleistocene epoch), while humans were evolving into their present form, the climate of Earth was swinging between long, cold glacial periods and brief, warm interglacial periods. Large ice sheets were growing and decaying. The level of the sea was rising and falling. Patterns of precipitation and vegetation were shifting. The climate records in deep sea cores and in ice cores show that 100- to 1,000-year swings dominated, with some significant fluctuations of shorter period as well. Embedded in the longer cycles were numerous brief events, such as the Heinrich and Dansgaard-Oeschger events and the dramatic Younger Dryas event at the end of the last glacial period, 12,000 years ago. In the Younger Dryas, the temperature may have jumped several degrees Celsius in a span of 20 to 50 years (NRC, 2002).

The implications of these recent natural climate changes for the modern world are profound, but the absence of a complete theory is unsettling. The longer cycles show a strong relationship between carbon dioxide concentration and climate, but without clear evidence of cause and effect. The brief embedded events demonstrate the ability of Earth's climate to shift quickly. Such sudden shifts today would have significant impacts on human life and well being.

Progress on a theory for the ice ages will require extensive field and laboratory work, along with the development and testing of numerical models. Ocean heat storage, cloud distribution, ice dynamics, and—possibly—atmospheric chemistry must be included in these models. Intensive use of global climate simulations will be called for, although they may not be a limiting resource.

Major Challenge 9: Model and Understand Key Climate Events in the Early History of Earth and Other Planets [3]

While the recent Pleistocene climate fluctuations were strong, Earth's earlier climate experienced even larger fluctuations. Examples include these:

- Paleocene-Eocene thermal maximum (55 million years ago), when ocean temperatures rose by about 6 degrees and the carbon cycle experienced a large anomaly.
- The Cretaceous-tertiary impact-induced extinction event (65 million years ago); the resulting changes in the environment that killed the dinosaurs remain a mystery.
- The Permian-Triassic extinction (250 to 290 million years ago), which decreased the diversity of life on Earth; the role of climate change is not known.
- Snowball Earth (about 700 million years ago); glacial deposits from that time suggest that the entire planet may have been ice-covered.

Just as an understanding of ancient terrestrial climates provides a context for modern climate studies, so does the understanding of the climate of nearby Earth-like planets. The most relevant planets in this regard are Venus and Mars. Venus is closer to the Sun than Earth is, has a much higher albedo, and a massive CO_2-rich greenhouse atmosphere. The surface temperature on Venus is nearly 700 K, twice that of Earth. The central question for Venus is whether it once had a cooler Earth-like climate and, if it did, how and when it developed its powerful greenhouse. Mars, on the other hand, is a darker planet, with a thinner CO_2 atmosphere and seasonal CO_2 ice caps. Its rotation rate is similar to Earth's, and the basic fluid dynamics of its atmosphere may resemble that of Earth. The main difference is the absence of large amounts of water, with the accompanying influence of latent heat. Together, the climatic history of early Venus, Earth, and Mars provides a useful test of our knowledge of climate change.

COMPUTATIONAL CHALLENGES IN THE ATMOSPHERIC SCIENCES

As noted in the committee's ratings for each of the major challenges, Major Challenges 1-3, 5, and 7 are critically dependent on advances in HECC capabilities, while HECC plays an important though probably not rate-limiting role for Major Challenges 4, 6, and 8. HECC plays a role in Major Challenge 9, but its absence would not be a barrier to progress.

HECC in the atmospheric sciences consists primarily of simulations based on coupled, multi-dimensional partial differential equation models of fluid dynamics and heat and mass transfer. These fundamental atmospheric processes are driven by a variety of forces produced by radiation, moisture processes, chemical reactions, and interactions with land and sea surfaces. The largest-scale flows in the atmosphere are a response to the poleward thermal gradients created by solar radiation falling on a spherical Earth. At these large scales, the dynamics of Earth's atmosphere are shaped by two effects. First, the vertical scales of structure and motion, which are small compared to horizontal scales, are essentially in hydrostatic balance. Second, the large-scale horizontal motions are largely constrained by the balance of Coriolis forces arising from Earth's rotation and horizontal pressure gradients. However, these large-scale balances are modulated by smaller-scale effects and other physical processes, which are reflected in the variations we observe as weather and climate.

The equations of fluid dynamics and heat and mass transfer are approximated by discrete forms in a fairly standard approach, with modifications to take into account the large aspect ratio of Earth's atmosphere and the range of length scales and timescales that must be represented. The remainder of this section describes the primary issues that distinguish high-end simulation in atmospheric science.

Chaos, Probability Forecasts, and Ensemble Modeling

The fundamental mathematical description of atmospheric dynamics as a well-posed initial-value approximation to the Navier-Stokes equations might lead one to believe that increasing the accuracy of observations and increasing the resolution of numerical forecasts would forever increase the accuracy of the numerical forecasts. However, this is not the case. It has been known since the 1960s (Lorenz, 1963; Smale, 1967) that solutions of otherwise well-posed nonlinear evolution equations can exhibit chaotic behavior, with exponential growth in small perturbations interacting with nonlinearity to produce incredibly complex behavior over long times. The chaotic behavior in fluid dynamics problems has been observed experimentally and in numerical simulations in a variety of settings, including atmospheric fluid flows. In a landmark study of errors in an operational weather forecast model, Lorenz (1982) reached the conclusion that the limit of predictability of weather events as an initial-value problem was about 2 weeks. Beyond that time period, the best one can expect is to be able to predict statistical averages.

As with the actual atmosphere, the fields generated from complex numerical models of the atmosphere are chaotic and apparently random. Simulations of the annual cycle of temperature and precipitation show a realistic variation from year to year, just like the actual atmosphere, but the superimposed randomness poses a challenge for modelers seeking to trace signs of climate change back to altered system parameters. Whether the change is due to a rise in carbon dioxide, a shift in ice albedo, an altered solar constant, or a change in aerosol concentration, it is necessary to simulate several years of behavior to gather statistically significant evidence of climate sensitivity. Among climate scientists, a minimum of 10 years of model integration is typically thought to be needed for credible work. Such a standard creates substantial demand for supercomputer power, a demand that compounds the needs arising from high spatial resolution and the addition of physical processes. The IPCC climate runs carried out by many research and operational centers used a substantial fraction of the available computer power in the atmospheric community from 2004 through 2006 to satisfy this requirement. Regional climate studies that nested within the global IPCC runs set 5 years as the credibility standard.

Today, the strategy for determining the likelihood of important events is to take advantage of increasing computer power to compute many forecasts in addition to single forecasts at higher resolution. These ensembles of forecasts are the critical strategy for coping with the implications of chaos. Usually, the ensemble is constructed by starting with a single forecast whose initial and boundary conditions and internal physics algorithms are the best available in some sense. This single forecast is referred to as deterministic and is often run at high resolution. The ensemble is then constructed by perturbing initial conditions, boundary conditions, and internal physics algorithms to produce a variety of forecasts starting from the same time. If the ensemble of forecasts appropriately reflects the uncertainty in the initial and boundary conditions and in the internal physical approximations, then the computed ensemble forecasts should provide a probability distribution that reveals the uncertainty in the forecast and that allows decisions to be made on a probabilistic basis.

Improvements to operational forecast capabilities require research into next-generation production systems. Those next-generation systems demand HECC almost by definition because they incorporate models and algorithms of greater complexity than those used in production systems, which themselves are heavy users of high-end capacity.

The major forecast centers all produce global ensemble forecasts extending up to 2 weeks into the future. For example, each day the ECMWF produces a global forecast with an ensemble of 51 members.

Palmer (2006) gives a specific example of how these forecasts can warn of extreme events. Increasing the resolution of the forecast and the number of ensemble members requires additional computer power. Because probabilities derived from sufficiently broad ensembles are so valuable, the general strategy today is to first set the number of ensemble members and then determine the resolution at which they can be run.

With the even greater uncertainties involved in numerical prediction of seasonal variations several months in advance, the ensemble strategy is essential and is used by the major centers to produce outlooks for a season or two in advance. The forecasts are often delivered as a mean anomaly from climatology, computed by averaging the ensemble of anomalies, or as a set of probabilities of being above or below normal, estimated by counting the number of ensemble members that fall into the appropriate categories.

The same strategy is used for long-term simulations of climate and global environmental conditions that are aimed at understanding paleoclimates or at predicting global conditions for decades or centuries ahead as the Earth system reacts to the forcing supplied by carbon dioxide released by combustion of fossil fuel. In seasonal and climate modeling, the interactions between the atmosphere, the ocean, and the land surface must be resolved and depicted accurately in order for the results to be reliable and representative.

For all timescales, the probabilistic forecasts are useful to decision makers in a wide variety of weather- and climate-dependent enterprises who use quantitative methods to distinguish opportunity and risk and to select specific courses of action.

Numerical Weather Prediction

The strategy for preparing weather forecasts to assist the public and private sectors with the management of weather risk has evolved dramatically during the era of numerical prediction models. Figure 3-2 provides a schematic of the overall process that leads to forecasts prepared by the U.S. National Weather Service's National Centers for Environmental Prediction (NCEP). The first U.S. forecast computed with the fundamental equations of motion became operational in the National Weather Service on June 6, 1966. At that time, the numerical forecast served to portray the large-scale evolution of the atmosphere, and human forecasters translated those patterns into weather events at the surface through theoretical reasoning and experience. The combination of improved observations, improved understanding of atmospheric processes, and higher-resolution models made possible by increasing computer power has produced increasingly skillful numerical forecasts over the past five decades. The output products have improved concomitantly, becoming more specific and more quantitative. A key improvement, and one still used today, was to develop regression equations between surface weather variables and variables in the numerical prediction by comparing forecasts with observations. Known as model output statistics (MOS), this system improves surface forecasts considerably when there is sufficient history that the regression equations at each forecast location can incorporate seasonal variation. Given their statistical structure, these are probabilistic forecasts even though they may not be stated in a probabilistic format.

There are many ways to assess the quality of weather forecasts (Wilks, 2005). Some concentrate on verification of forecast products with surface observations; others are designed to assess how well the numerical forecasts portray the large-scale features in the middle atmosphere that control the evolution of the surface weather. The committee chose the second option and looked at the S1 skill score (Teweles and Wobus, 1954), which compares predicted gradients in the middle of the atmosphere with those actually

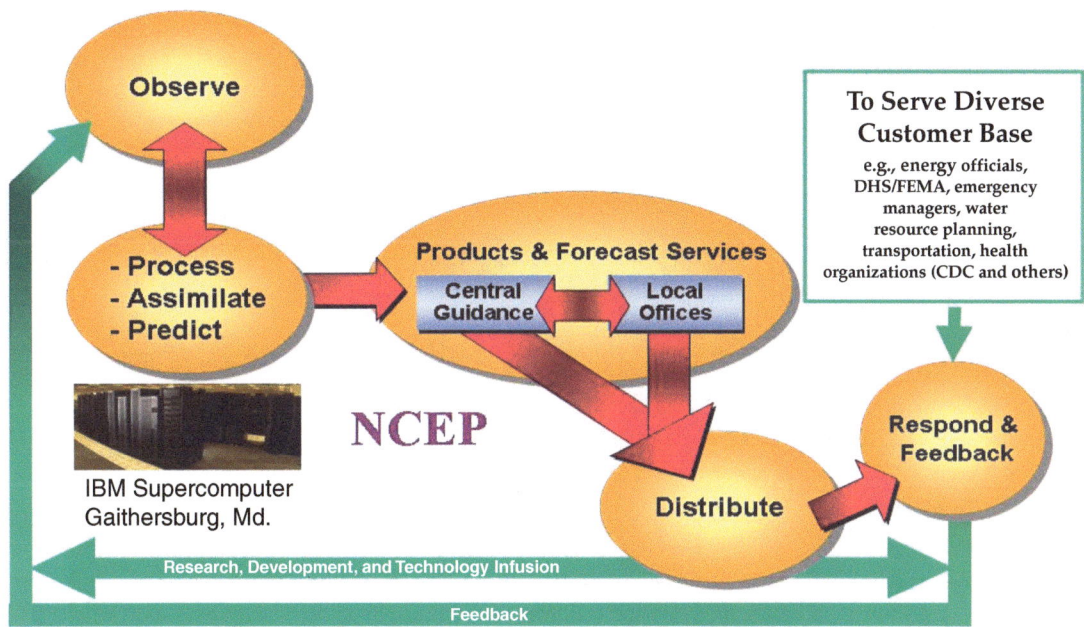

Prediction is now inherently linked to numerical models

FIGURE 3-2 Schematic of the process for preparing National Centers for Environmental Prediction forecasts.

observed. Using gradients to measure skill is a demanding but valuable approach because the models must handle gradients well in order to depict strong weather systems and strong winds accurately.

The increase in skill of the numerical models of the National Weather Service over the past 50 years is shown in Figure 3-3. The skill score is derived from the S1 score, as indicated in the figure. Note that the 72-hr forecast today is as skillful as the 36-hr forecast was some 20 years ago. A wide range of technological and scientific advances are responsible for this improvement in numerical weather prediction. Key factors are better observations from meteorological satellites (as spatial, temporal, and spectral resolution improved), advances in numerical representation of small-scale atmospheric processes, including precipitation mechanisms, and greater spatial resolution of the prediction models themselves. In each of these, the dramatic increase in computer power over the past few decades played an essential role.

The measurement of forecast skill for large-scale phenomena is easier than measuring skill for very local events, including precipitation, cloudiness, and wind gusts. Such events occur at much smaller scales than the scales being used in the computation and, at present, can only be represented statistically. Over the history of numerical weather prediction, forecasts of local events have improved in parallel with forecasts of more widespread conditions. Very-high-resolution models and computations offer the possibility of significant improvement as important aspects of atmospheric flows are included directly in the computation. The numerical computation of a weather or climate forecast must start from an initial condition representing the state of the atmosphere at the time the forecast begins. The more comprehen-

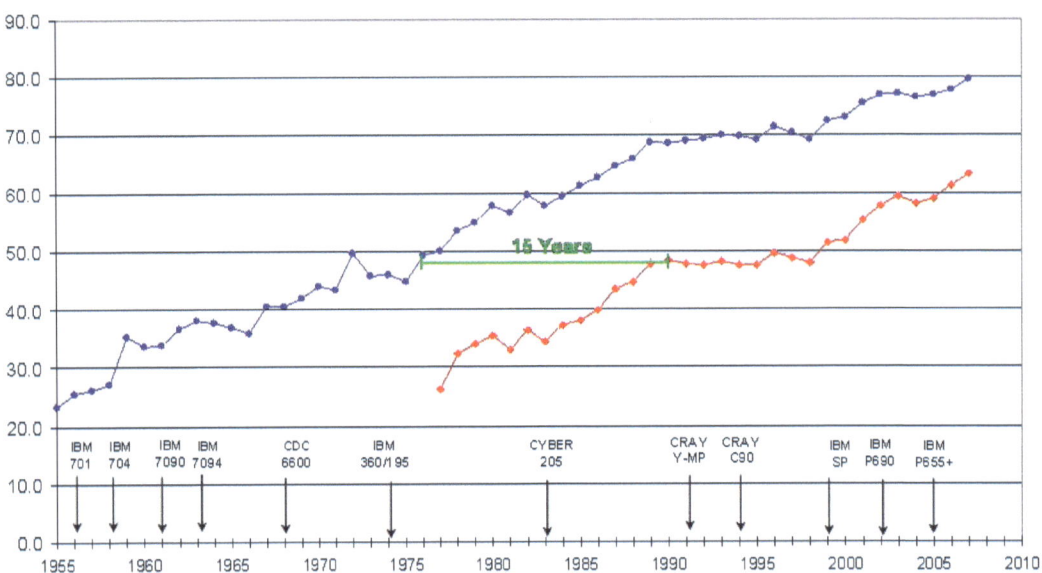

FIGURE 3-3 Skill of the National Weather Service numerical forecast, derived from the S1 skill score discussed in the text. This progress is due to number of factors, including faster computers that permit computation of the forecasts at higher resolution. Using data from the NCEP on 36- and 72-hr forecasts at 500 mbar over North America, the figure shows that in recent years the accuracy of forecasts has increased faster than would have been expected based solely on the finer resolution made possible by hardware upgrades: Whereas spatial resolution improves as the 1/3 power of computer capability, these curves grow as the 1/2.2 power of computer capability. This demonstrates the payoff for balanced investments in machines and software. SOURCE: NCEP Central Operations, January 2008.

sive and more accurate the initial representation is, the better the forecast will usually be. We see here an immediate problem: observations obtained from surface and balloon-borne instruments are plentiful over the developed regions of the world but are scarce, at best, over the rest of the world, including the oceans. Thus one of the main roles of Earth-sensing satellites in polar and geosynchronous orbits is to obtain global observations of the atmosphere, ocean, and land that will define atmospheric conditions for the initialization of numerical forecasts. The assimilation of satellite data has increased dramatically over the past decade, not only in quantity but also in the range of characteristics measured, with data from satellites now exceeding data from conventional observations by factors of 10 or more.

The preparation of initial conditions is itself a complex task (see, for example, Kalnay, 2003). Forecasters must take observations of varying quality and at different scales from all over the world and combine them into an integrated numerical picture of the state of the atmosphere. This is a delicate endeavor, because if the initial condition is not dynamically consistent so that the motion and thermal fields balance, the model will start with strong accelerations and rates of change, generating a balanced but perhaps unrepresentative state at the beginning of the forecast. Typical of the current state of the art is the use of four-dimensional variational assimilation, or 4DVAR, which runs over the three-dimensional spatial domain for an interval of time seeking to converge to evolving fields that are dynamically consistent and as close as possible to the original observations. This can be an ongoing process: While a prediction run started from a 4DVAR assimilation proceeds over a period of hours, another 4DVAR process can be started so as to be ready for the next prediction cycle. In effect, the model is run continuously to assimilate observations and deliver an initial state for the next prediction run. Such data assimilation schemes are computationally expensive. As satellite observations make use of increasing numbers of spectral channels and deliver data at significantly higher rates, the assimilation computations will pose computational requirements equal to those of the prediction models themselves. This is discussed further below.

Climate Modeling

Climate modeling has evolved into a solidly physics-based science of planetary atmospheres that relies on classical atmospheric physics, radiation, and chemistry and incorporates extensive interactions and feedbacks between the atmosphere, land surface, and oceans, with their inherent chemistry and biology. These models simulate the physical, chemical, and biogeochemical Earth system and its environment in an interactive, feedback-based methodology. The numerical representations may incorporate subsystems that model the impacts of climate change on plants, animals, and people and for simulating adaptation and mitigation strategies. Past and present ocean-atmosphere general circulation models were designed primarily for physical, chemical, and biological scientific applications. Earth system models of the future will be designed to support social decision making as well. They will simulate weather and climate from the local scale on up to the global scale and will be coupled with models of ecosystems and of socioeconomic policy.

The propagation of uncertainty and variability through coupled physical-biogeochemical models is challenging, particularly because nonlinear interactions can amplify the forced response of a system. New systematic theories related to multiscale, multiphysics couplings are required to better quantify feedbacks and relationships. This is critically important as global integrated Earth system model results are used to force, couple, and tune economic and impact models. Groundwork on the science of those relationships and quantification of uncertainties through coupled biogeochemical, physical, and economic systems are necessary to support the complex decisions that will be made over the coming decades.

Threat-Focused Operational Forecasting

Contemporary numerical weather forecasts in both deterministic and ensemble form usually succeed in anticipating threatening weather conditions, but they do not have adequate internal resolution or resolution of Earth's surface to provide sufficient information on the track and intensity of threatening storms or of local effects. When satellite and other observations and numerical forecasts show tropical storms or hurricanes on trajectories that may approach the United States, special models are run to obtain more information on the time and place of landfall and the intensity of the storm as it nears the coast.

The efficacy of such high-resolution modeling is exemplified by the simulation of Hurricane Katrina with the community's high-resolution Weather Research and Forecast (WRF) model nested at resolutions of 36, 12, and, finally, 4 km in the U.S. Global Forecast Model. Figure 3-4 shows the simulated surface pressure pattern as the center of the storm made landfall; it illustrates that the track of the hurricane was computed with notable accuracy. The figure also shows a simulated radar image derived from the distributions of water and ice computed in the simulation model.

Contemporary high-resolution simulations confirm that the statistical subgrid-scale algorithms now used in atmospheric models are generally adequate down to horizontal scales of about 10 kilometers. However, one of the interesting features of simulations such as the one for Hurricane Katrina is the

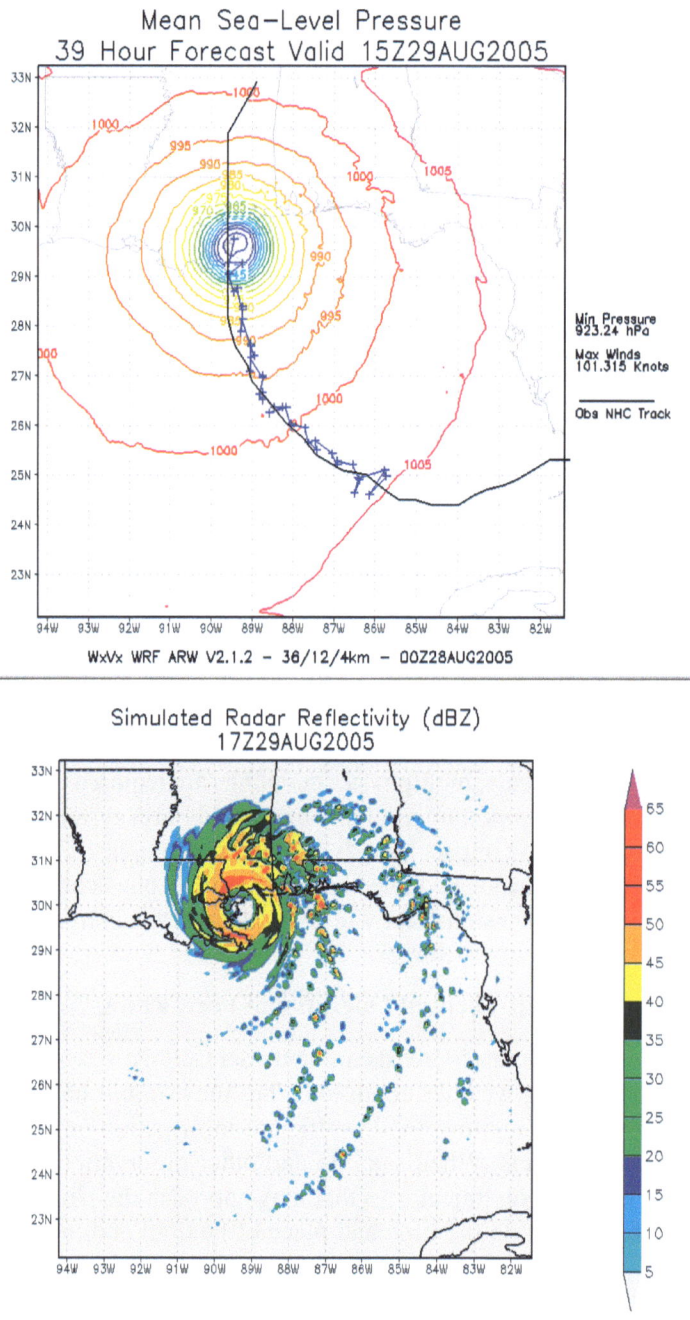

FIGURE 3-4 Simulation of Hurricane Katrina using the Weather Research and Forecast (WRF) model at 3 km resolution obtained by a series of nested computations based on the Global Forecast System. The upper panel shows the surface isobars at landfall and the simulated track of the hurricane as blue arrows compared with the observed track in black. The lower panel shows a simulated radar image of the rainfall in the hurricane at landfall computed from the distributions of water and ice in the simulated storm. Computations performed by Weather Ventures Ltd.

significant increase in realism that occurs at resolutions of 4 km or less when the convective processes are calculated directly instead of with the statistical schemes that are used at 12 or 36 km resolution. Statistical algorithms to represent some subgrid-scale processes will also be required, so work on them continues.

Threat-based operational forecasting is possible for a variety of other threats, but it is not generally implemented today. Thus there is a significant opportunity for such forecasts to predict at high resolution impending snowstorms, severe weather complexes, serious air quality events, and emergency situations such as wildfires, release of toxic plumes, or severe icing events.

Implementing threat-focused, high-resolution forecasts poses a number of challenges. The first is identifying the cases that merit such attention. The second is accessing the substantial computer power necessary to make such forecasts in a short period of time, because the supercomputers that are capable of such a targeted calculation are generally running full-time to produce normal operational forecasts. The third is identifying members of the large-scale ensemble to use as the foundation for the special forecast. With observational and forecast systems providing increased rates of information flow in the years ahead, we can anticipate that threat-focused forecasting will become more valuable and will add to the demand for computer capability in numerical weather prediction.

Spatial Resolution, Adaptive Grids, and Subgrid Processes

The physical processes that determine the dynamics of Earth's atmosphere occur over an enormous range of scales: spatial scales from microns (for water vapor and ice) to thousands of kilometers, and timescales from seconds to centuries. No computer simulation now or in the foreseeable future will be able to resolve the full range of spatial scales and timescales. Typically, simulations are set up to represent some fraction of the timescales and length scales, from the largest down to some smaller limit of resolution, with processes that take place at still smaller scales represented using so-called subgrid-scale models. Examples of subgrid-scale models incorporated in atmospheric simulations include the following:

- *Convective processes*. The convective components of numerical models include relatively intense vertical motion driven by buoyancy and often leading to cloud formation, at scales that are generally too small to be resolved by the simulation's spatial grid. Thus schemes must be introduced to represent condensation and latent heat release; large-scale motions within the clouds and the exchange of air with the region near the cloud; and the mass fluxes of air and the three phases of water.
- *Cloud microphysics*. All of the complex interactions involving water in the cloud must be modeled to account for water vapor, cloud water, cloud ice, rain, snow, graupel, and hail. The rates at which raindrops and ice crystals grow must be estimated and their fall velocities calculated. Occurring on scales from the molecular to a few centimeters, these processes must be represented statistically in all atmospheric models.
- *Radiation*. Radiation from the Sun warms Earth and its atmosphere while the emission of long-wave radiation from the top of the atmosphere maintains its thermal balance. The scattering and conversion of radiation into other forms of energy as it travels through the atmosphere also occur on scales so small that statistical models will be required for the foreseeable future. The models of radiation processes must take account of wavelength dependencies, and they are often based on strategies that involve treating spectral variation in bands. Detailed understanding of

how atmospheric molecules accrete to form aerosol particles is also critical for understanding atmospheric radiation, and it must be represented by a subgrid-scale model.
- *Boundary layer processes.* Atmospheric flow is profoundly affected by conditions at the surface of the Earth, with thermal exchanges and the loss of momentum creating a boundary layer that may extend 3 or 4 km in the vertical. Turbulent motion is responsible for much of the transfer of heat and water in the boundary layer. The vertical resolution of contemporary global models is higher in the boundary layer and is often on the scale of a few hundred meters. The most sophisticated statistical schemes for portraying boundary layer dynamics are now developed in very high-resolution models that produce large-eddy simulations. These statistical schemes can be used for weather prediction and research models on scales of a few kilometers or more.
- *Surface layer.* The actual exchanges of heat, water, and momentum between the atmosphere and Earth occur in a surface layer usually only tens of meters deep at the bottom of the boundary layer. Here the mechanical generation of turbulence owing to wind shear is dominant, and the fluxes of energy are strong. Over the ocean, we must model the interaction of two fluids at a common boundary; over the land, we must take account of topography, land use, running water, and melting snow, as well as conditions in the upper layers of the ground. Thus, contemporary prediction and research models are linked to models of the ocean and the land surface, which themselves are complex, dynamical models that compute the evolution of a wide variety of processes at a wide variety of scales. But the actual exchange between the atmosphere and the ocean and land surface occurs on scales much smaller than those in these models, so it must be modeled statistically on both sides of the interface.

Very-high-resolution research simulations of these processes, using detailed physical and chemical equations, provide rich data sets for developing algorithms whose implications can then be tested by comparison to appropriate observations and by measuring improvements in large-scale simulation models.

Preserving, Improving, and Using Weather and Climate Observations

The store of atmospheric and environmental observations is a continually expanding resource for research and operations as the new data each day add to our understanding of Earth. But this resource is more than a static archive, because continuing progress in scientific understanding and technological capability makes it possible to upgrade the quality and applicability of the stored data when computer resources permit.

In recent years, satellite data have been playing an increasingly important role in weather forecasting as the primary forecast centers process data flows that go well beyond 1 billion observations per day. Satellite observations of winds, temperatures, and humidity over the world oceans, previously a region devoid of data, are contributing greatly to this flow. Some of the difficulties in merging different types of data taken over a period of several hours have been solved using four-dimensional data-assimilation schemes.

Information from satellites is also being used to detect weather and climate phenomena such as shrinking glaciers and sea ice, advancing deserts, shifting ocean currents, and wide-ranging dust storms. This information comes from nearly 50 environmental satellites, providing a daunting flow of data. The distribution and analysis of these data are largely limited by the availability of large disk arrays and high-speed computers.

In the same way we use an atmospheric prediction model to assimilate data for a forecast run, we can improve historical observations through a computational process known as reanalysis. The reanalysis process can add variables not originally observed, and it can produce significant improvements in the reanalyzed data set obtained as the forecast models improve. Today there are a number of global, long-term reanalysis data sets available for research purposes and for applications. A notable data set just completed is the North American Regional Reanalysis (NARR) (Mesinger, 2006), which contains a wide variety of atmospheric and surface variables at a resolution of 32 km for the period from 1979 to the present.

These data sets are useful for developing dynamically consistent climatologies and for serving as verification data sets for calibrating ensemble models. Moreover, they provide a dynamically consistent set of initial conditions and verification data for use in model development and refinement. They offer, in effect, a balanced and consistent virtual atmosphere, but they are expensive because they require running a forecast model over many years of observations.

Reprocessing is also important with satellite observations, because continuing research often provides improved algorithms for converting the radiances observed by the satellite instruments into physical variables in the atmosphere or on the surface. Generally, the most basic information from the satellite instruments is securely archived so that this process can be carried out when the considerable requirements for computer capacity can be met. As an example of the computational challenge associated with massive satellite data sets, consider the reprocessing of global data from the Moderate Resolution Imaging Spectrometer (MODIS) satellite-borne sensor. Since the launches in 1999 and 2002, the research community has improved the algorithms used to derive physical climate variables from the satellite data (Vermote et al., 2002). With each improvement, the entire historical MODIS data set must be reprocessed. With current NASA computing capability, about 4 days of data can be processed in 1 day of computation. Thus, to generate a new historical MODIS product suite requires 1 or 2 years of dedicated computation. It is quickly becoming clear that, with the massive amounts of data generated by existing Earth observation systems, HECC is necessary for managing and analyzing data.

Atmospheric models, like scientific instruments, must be calibrated, especially if they are to be used for long-term prediction. A model intended for climate research, perhaps on global change issues, must be calibrated to produce the present climate by comparing key averages with observed equivalents, perhaps from a reanalysis data set. The calibration often requires adjusting parameters in the statistical algorithms for subgrid-scale processes or in the schemes controlling fluxes between the atmosphere, ocean, and land surface.

Similarly, the characteristics of models used to predict seasonal climate variations must be determined so the predictions can be corrected. A first concern is that such models will exhibit spatially dependent biases—for example, tending to be too warm in one part of the world and too cool in another. Typically, they are used to make retrospective forecasts for 25 years or so in order to determine the biases with sufficient confidence to correct the predictions.

A second concern with ensemble models is the spread of the members of the ensemble. A set of retrospective forecasts and verification data over an equal number of years is required to perform the statistical analyses needed to improve or calibrate the predictions.

The necessity of computing retrospective forecasts over a period of 25 years before bringing a new seasonal prediction model into operational service is a serious impediment to incorporating improvements into the computer prediction system or using observations. A common strategy is to compute ensembles with fewer members than will be used in practice, which saves machine time but limits statistical confidence.

HECC has a key role in enabling us to capitalize on all these terabytes or more of data. In many problems of atmospheric science, the greatest computational challenges (and number of operations) are for data assimilation, not modeling. Experts in the field told the committee that just handling the data volumes projected from satellites of the National Polar-orbiting Operational Environmental Satellite

System (NPOESS) era will necessitate a 10-fold or so increase over today's HECC capabilities. But other observational capabilities are also ramping up, such as those for terrestrial biogeochemistry. The Orbiting Carbon Observatory (OCO), for instance, is expected to be launched in 2008 and to generate more data than can ever be assimilated into today's models on today's computers. The computational attributes of assimilation are much different from those of modeling. For instance, the mix of serial and parallel codes is different. And the data treatment and assimilation part of the problem call for compromises to make the problem tractable and to get satisfactory data-usage numbers. These compromises have more serious consequences than the compromises in modeling. Some would even argue that the potential to improve weather forecasting by improving assimilation is significantly greater than the potential by improving modeling.

The computational environments needed for weather and climate studies are strongly influenced by the observational data. The size of the data sets is important, but maybe more important is the growing complexity of the data system. Further, because there are real-world uses of weather and climate products, those products are subject to forms of analysis, evaluation, and validation that are far different from those used for the products of most other fields of science and engineering. This evaluation is data driven in a way that is not defined by the normal practice of "science"; for instance, the performance of a global model that incorporates more than 1 billion observations might essentially be evaluated by how well it predicts a snowfall in one evaluator's home town.[1]

The assimilation of data into predictive models, as well as the new understanding of the atmospheric system that the combination of observations and modeling will produce, is an exciting development in Earth system science. We will need more computing capability to handle the higher-resolution, higher-fidelity measurements. The ability to efficiently explore data sets that will be produced by the anticipated high-resolution models is not yet clear, nor is the path to analysis and visualization of those vast quantities of data. These activities are absolutely essential to the research enterprise. Equivalent increases in storage and archive capacity will also be required, as well as additional related expertise in software engineering and computer science.

Transitioning to New HECC Resources

The atmospheric sciences are ready to exploit additional HECC resources right now. Some coupled climate simulations have been performed utilizing up to 7,700 processors. If more resources were available, current models could be run more routinely on such platforms, enabling the incorporation of more, and more-realistic, physics and chemistry. It is expected that the scientific benefit would be immediate. In operational meteorology, a 10-fold improvement in computing capability could readily be exploited in the near term, noticeably improving the accuracy and quality of weather and ocean predictions. Increases of this order are required to improve prediction of severe weather, as well as for forecasts of importance to the energy, transportation, and agriculture sectors. The operational forecast centers could become ready to use the increased resolution during the interval between the purchase decision and delivery of the new computer.

Further advances, though, would require modifying the current software to improve scalability and changing the model physics to reflect phenomena that become important because of finer resolution. That step would be followed by a significant expenditure of time and computational resources to establish the baseline forecast performance of the models over enough cases to reflect a reasonable range of weather conditions.

---

[1]The committee thanks an anonymous reviewer for many of these thoughts on data assimilation.

More generally, the HECC requirements for atmospheric science must consider the overall end-to-end science enterprise. While the Japanese Earth Simulator has done wonders for energizing the climate science community in Japan and has been extremely useful for collaborators worldwide, there was no way to use the computer except by going to Japan. While grid computing cannot help with high-end processing, it can help with access, scheduling, monitoring, and analysis of simulation results. This analysis step is—next to developing the models themselves—the endeavor where scientific judgments play the largest role. Indeed, the infrastructure for analysis and access to simulation results is a key part of the computational modeling in the atmospheric sciences.

For weather forecasting, the entire end-to-end system is executed in a short number of hours, and success might be evaluated in less than 48 hr. This time element is another critical, defining characteristic of computing for operational weather forecasting. The data capture, quality control, assimilation, and forecasting take place in a few hours, and are repeated every 6 hr. This throughput requirement drives the capability requirements of the computing system. Along with the consequential and tangible nature of the products, it also calls for a hardening of the computing system that is far more stringent than called for by most scientific research. This brings attention, again, to the end-to-end system, the stability of the hardware, and the full range of systems software that glues the system together.[2]

In any HECC upgrades, software needs continual adaptation. There may be significant rewrites ahead of the programming paradigm changes from the MPI and OpenMP language extensions. The HPC language program of the Defense Advanced Research Projects Agency (DARPA) and the special-purpose (game) processors that are starting to encroach on computer offerings are likely to have a large impact on the effort required to stay with the technology curve. Already, model development calls for equal investments in software engineering and science. While, as noted above, some additional HECC capacity could be readily absorbed by the atmospheric sciences, the longer-term science goals—creating the capability for regional climate forecasts and for localized weather forecasts that explicitly model convective processes in enough detail to predict severe weather—will require significant attention to the overall system. In particular, there will be a need for modifications to operational modeling systems and significant changes in physical parameterizations at all scales.

THE NEED FOR HECC RESOURCES TO ADVANCE THE ATMOSPHERIC SCIENCES

For nearly 60 years, the atmospheric sciences have utilized the most powerful HECC resources that were available, starting with the first numerical weather forecast computed on the ENIAC in 1950.[3] Today the most advanced computers are used in predicting daily weather and seasonal climate variations, in examining likely scenarios for climate change over the remainder of this century, and in a variety of research efforts aimed at understanding atmospheric processes and phenomena and their interactions with other components of the Earth system. The full coupling of all the factors that influence climate, atmospheric science, and weather requires vast computational resources. The computational challenges and strategies are diverse, and new ones now arise as observations of the Earth and atmosphere increase in number and diversity.

There are three areas suggested by this chapter where advances in HECC would speed up the achievement of important scientific and national goals, each of which is described on the next two pages.

---

[2]The committee thanks an anonymous reviewer for contributing the ideas behind this paragraph.
[3]Platzman (1979) gives a detailed account of the early days of numerical forecasting.

Improved Models for the Global and Regional Hydrologic Cycle

A variety of phenomenological macroscopic models are currently used to represent clouds, precipitation, and convection in both weather forecasting and climate modeling. However, they give widely varying results when applied to the same problems. Furthermore, they are not valid for grid resolutions finer than about 10 km. Finally, it is not known how to make a direct connection between this class of models and detailed mesoscale and microscale observational data.

It would be possible to use more realistic models of clouds, precipitation, and convective processes by substantially increasing the grid resolution, which would be feasible with enough computing power. At a horizontal mesh spacing of 1 km, it is possible to replace cumulus convection and large-scale precipitation parameterizations with a new class of models that resolve cloud systems. Such fine-grid resolution would also permit the accurate prediction of tropical cyclones and other extreme weather events. The difficulty with such an approach is computational expense: It has been estimated that a global model on a uniform grid with 1 km mesh spacing would require a computer with sustained performance of 10 Pflop/s (Oliker et. al., 2006), which would require parallel performance and scalability orders of magnitude beyond current capabilities. However, these issues are being addressed, and techniques of computation and scaling are evolving along with hardware capabilities.

An alternative approach that would require far less computational power is adaptive mesh refinement (AMR), in which grid resolution can change as a function of space, time, and the emerging solution (Berger and Oliger, 1984). Various kinds of nested grid methods closely related to this approach have been used in atmospheric modeling over the past 20 years, as illustrated in Figure 3-4. Most of these methods are based on one-way coupling, in which a solution computed on a coarser global grid is used to interpolate boundary conditions for a fixed, nested finer grid, with no feedback from the fine-grid solution to the coarse-grid evolution. The use of AMR requires a two-way coupling between coarser and finer meshes, which raises considerable theoretical and practical difficulties (Jablonowski, 2004), from fundamental issues of well-posedness (Oliger and Sundstrom, 1978) to an observed tendency of local refinement methods to produce unphysical precipitation at refinement boundaries.

A successful attack on these issues should probably pursue both approaches: the development of (1) scalable implementations of uniform-grid methods aimed at the very highest performance and (2) a new generation of local refinement methods and codes for atmospheric, oceanic, and land modeling. Both efforts should be undertaken with the specific goal of improved models of the global and regional hydrologic cycle, including validation against detailed observational data.

Better Theory for and Quantification of Uncertainty

Atmospheric models are exceedingly complex owing to the multiple physical, chemical, and biological processes being represented, the numerical algorithms used to represent these processes individually, and the coupling of the individual processes at the level of both the mathematical models and the numerical methods. The propagation of uncertainty through a coupled model is particularly problematic, because nonlinear interactions can amplify the forced response of a system. In addition, it is often the case that we are interested in bounding the uncertainties in predictions of extreme, and hence rare, events, requiring a rather different set of statistical tools than those to study means and variances of large ensembles. New systematic theories about multiscale, multiphysics couplings are needed to quantify relationships better. This will be important as atmospheric modeling results are coupled with economic and impact models.

Building a better understanding of coupling and the quantification of uncertainties through coupled systems is necessary groundwork for supporting the decisions that will be made based on modeling results.

This understanding requires new approaches to error attribution and a better mathematical theory for complex model systems. The result would be models that are demonstrably more accurate. One aspect of the challenge is that error quantification and propagation have to span a dozen or more orders of magnitude just for the time component—ranging from timescales of 10^{-6} s for some chemical reactions in the atmosphere up to 1,000 years for the characteristic timescale of ocean circulation. If we understand the sources of uncertainty, we may be able to make models so good in particular areas that we are at the limits of what we can learn from observation. Conversely, where models are uncertain, we may be able to suggest observations or experiments that would significantly add to our knowledge of the climate system.

Continued Development of HECC Infrastructure

The atmospheric sciences have been relatively successful at developing community infrastructure for HECC. This is largely due to the programmatic requirements of operational weather prediction and, in climate modeling, of the IPCC assessment process. Components of this infrastructure include widely available and well-supported community codes with well-defined interfaces to many of the physical submodels that permit experimentation by the community; access to HECC computing systems that enable scientific investigation using simulation; and standards for archiving and sharing simulation data. As noted in the section "Computational Challenges in the Atmospheric Sciences," valuable scientific and operational advances would be within reach of the community if additional HECC resources, with computing power up to 10 times greater than that in use, were made available. These advances, which would bring more valuable operational forecasts and better understanding of feedbacks and coupling in Earth's climate, are achievable with the community's current structures.

Longer term, however, computational capabilities will need to increase 1,000- to 10,000-fold. Such an increase would not be possible with the current infrastructure. The increases in code complexity could exceed the capacity of the national centers for software development and support, while the growth in model complexity and the need for better resolution would greatly increase the need for computing cycles. Moreover, the data requirements before and after model simulation or prediction threaten to swamp the system. The volumes of data from simulations and from satellite and other observations are rapidly increasing, and the reanalysis of that data for the purpose of scientific investigations and policy analyses will stress the infrastructure for storing and accessing the data, necessitating a whole new generation of analytical tools.

The solution to these problems lies in a combination of hardware, software, and human resources. The national centers and the federal agencies that fund them already have plans for increasing the capability of computing facilities to a petaflop and beyond, so that the main concern is that the atmospheric sciences community be able to access these resources. Some of the computational needs related to data management and analysis are already acute, and the scalability, algorithms, and software hurdles will become paramount when we begin to move to platforms with multiple processors per chip, whether or not the platform has petascale performance. Historically, the community has usually been successful in obtaining HECC resources for the more programmatic activities, but its track record is not perfect, and there must also be sufficient access for the research community. Software and human resources represent a more difficult problem. To overcome the challenges, it will be necessary to develop production-quality versions of a whole new generation of simulation codes, analysis codes, and middleware for managing the data, while at the same time maintaining and enhancing the current capabilities to meet evolving atmospheric science requirements. Such an effort will require the formation of focused development teams with close ties to the algorithm and software development communities on one side, and to the science users and observational programs on the other.

CONCLUSION: EARTH IN A COMPUTER

This chapter has shown how the atmospheric sciences combine global observations from Earth's surface and from space with scientific algorithms, software systems, and high-end computers to predict weather and seasonal variations and to simulate long-term climate change. The computer models and their predictions are the focus of a widely distributed effort to understand and resolve the complexity of the atmosphere and the Earth system and to foretell their future. From weather events in the next hour to climate variations in the next century, these computer predictions and simulations are vital to a wide range of public and private activities involving issues of economics, policy, and sustainability.

A number of trends are rapidly increasing the demand for computational capability in the atmospheric sciences. First, the dramatic increases in the volume and resolution of observations from the surface and especially from space will necessitate computational power approaching that needed by the models themselves. Second, the incontrovertible evidence that forecasts improve with higher spatial resolution calls for geometric increases in computer capability. Third, the necessity of including and resolving the biogeochemical component of the Earth system in climate studies adds another dimension to the demand for computer power. These demands are not being met by existing computational resources.

The steady progress over the past 50 years has been impressive—each increase in computer capability has produced meaningful improvement in the value of the forecasts and simulations and, in turn, posed new scientific and algorithmic challenges. This strong record of success suggests that a leap now to petascale computational power would make further significant progress possible.

A virtual Earth system (VES) might run continuously in linked petascale machines, assimilating data from the tens of satellites in space and from nodes acquiring local data all over the world. VES[4] would maintain a continuous, dynamically consistent portrait of the atmosphere, oceans, and land—it would be a digital mirror reflecting events all over the planet. The image in that digital mirror would serve as the foundation for predictive models to run continuously, predicting global-scale weather and global-scale seasonal change at unprecedented resolution and with remarkable verisimilitude. These global portraits, in turn, would be the evolving matrix in which nested and focused high-resolution models sharpen the forecasts of the atmospheric events that really matter to humans and their societies.

The new dynamic record of Earth and the predictions of the VES model would bring forth an era of enlightened management of weather and climate risk, contributing to national economic vitality and stimulating a stronger commitment to environmental stewardship. The creation and operation of an accurate and reliable VES would be a stunning and commanding national achievement—a dramatic demonstration of the benefits that can be realized for society by linking Earth and atmospheric science with the most advanced computers.

REFERENCES

Cox, P.M., R.A. Betts, C.D. Jones, S.A. Spall, and I.J. Totterdell. 2000. Acceleration of global warming due to carbon-cycle feedbacks in a coupled climate model. *Nature* 208: 184-187.

Dutton, John A. 2002. *The Ceaseless Wind: An Introduction to the Theory of Atmospheric Motion.* New York, N.Y.: Dover.

Hack, James J., Julie M. Caron, G. Danabasoglu, Keith W. Oleson, Cecilia Bitz, and John E. Truesdale. 2006. CCSM–CAM3 climate simulation sensitivity to changes in horizontal resolution. *Journal of Climate* 19(11): 2267-2289.

Intergovernmental Panel on Climate Change (IPCC). 2001. *Third Assessment Report, Climate Change 2001.* Cambridge, England: Cambridge University Press.

---

[4]In a pleasant coincidence, these are the initials of the late Verner E. Suomi, the father of satellite meteorology, an incredibly innovative scientist, and a leader in creating effective communication systems to empower meteorologists to share meteorological information and advancing technology.

IPCC. 2007. *Fourth Assessment Report, Climate Change 2007.* Cambridge, England: Cambridge University Press.

Jochum, Markus, Raghu Murtugudde, Raffaele Ferrari, and Paola Malanotte-Rizzoli. 2005. The impact of horizontal resolution on the tropical heat budget in an atlantic ocean model. *Journal of Climate* 18(6): 841-851.

Kalnay, Eugenia. 2003. *Atmospheric Modeling, Data Assimilation and Predictability.* Cambridge, England: Cambridge University Press.

Lorenz, E.N. 1982. Atmospheric predictability experiments with a large numerical model. *Tellus* 34: 505-513.

Mesinger, Fedor, Geoff DiMego, Eugenia Kalnay, et al. 2006. North American regional reanalysis. *Bulletin of the American Meteorological Society* 87: 343-360.

National Research Council (NRC). 2002. *Abrupt Climate Change: Inevitable Surprises.* Washington, D.C.: The National Academies Press.

Palmer, T.N. 2006. Predictability of weather and climate: From theory to practice. In *Predictability of Weather and Climate.* Tim Palmer and Renate Hagaedorn, eds. Cambridge, England: Cambridge University Press.

Platzman, George W. 1979. The ENIAC computations of 1950—Gateway to numerical weather prediction. *Bulletin of the American Meteorological Society* 60: 302-312.

Rojas, Maisa. 2006. Multiply nested regional climate simulation for southern South America: Sensitivity to model resolution. *Monthly Weather Review* 134(8): 2208-2223.

Stephens, G.L. 2005. Cloud feedbacks in the climate system: A critical review. *Journal of Climate* 18: 237-273.

Teweles, Sidney, Jr., Herman B. Wobus. 1954. Verification of prognostic charts. *Bulletin of the American Meteorological Society* 35: 455-463.

Thompson, S., B. Govindasamy, A. Mirin, K. Caldeira, C. Delire, J. Milovich, M. Wickett, and D.J. Erickson III. 2004. Quantifying the effects of CO_2-fertilized vegetation on future global climate and carbon dynamics. *Geophysical Research Letters* 31.

Vermote, E.F., N.Z. El Saleous, and C.O. Justice. 2002. Atmospheric correction of MODIS data in the visible to middle infrared—First results. *Remote Sensing Environment* 83: 97-111.

Wilks, Daniel S. 2005. *Statistical Methods in Atmospheric Science.* New York, N.Y.: Academic Press.

4

The Potential Impact of HECC in Evolutionary Biology

INTRODUCTION

The dictum of Theodosius Dobzhansky (1964)—"nothing makes sense in biology except in the light of evolution"—has never been truer than it is today. With the rise of such fields as comparative genomics and bioinformatics, evolutionary developmental biology, and the expanded effort to build the tree of life, the discipline of biology has become increasingly dependent on the inferences, methods, and tools of evolutionary biology. These contributions from evolutionary biology have become standard in solving problems in comparative biology, the biomedical and applied sciences, agriculture and resource management, and biosecurity. Thus an understanding of evolutionary change in individuals and populations provides the foundation for advances in crop improvement and vaccines, for improved understanding of epidemiology and antibiotic resistance, and for managing threatened and endangered species, to name only a few (Meagher and Futuyma, 2001).

Likewise, at the level of species and multispecies lineages, our new understanding of the tree of life is providing a comparative framework for interpreting the similarities and differences among organisms. Using newly generated knowledge about the phylogenetic relationships of life on Earth, comparative biologists have been able to do interesting and useful things:

- Identify wild relatives of domesticated plants and animals, leading to enhanced food security.
- Create tools for the discovery of countless new life forms, many of which are economically important.
- Establish a framework for comparative genomics and developmental biology, which speeds up the identification of emerging diseases (such as avian influenza and West Nile virus) and helps to locate their places of origin.
- Advance such disparate fields as resource management and forensics (Cracraft et al., 2002).

The committee was charged with reviewing the important scientific questions and technological problems in evolutionary biology by drawing on survey documents. Unlike the other fields covered in this report—astrophysics, the atmospheric sciences, and chemical separations—evolutionary biology has no definitive reports setting forth its scientific breadth and describing future challenges. Thus in developing its evaluation of the main scientific challenges facing evolutionary biology, the committee drew on workshop reports, primarily those prepared for the National Science Foundation (NSF, 1998, 2005a, 2005b, 2006), and on a document produced by eight scientific societies (Meagher and Futuyma, 2001), as well as on discussions among committee members and invited experts at a small workshop in December 2006, the agenda of which is included in Appendix B.

This chapter identifies the main challenges of evolutionary biology and evaluates the extent to which computational methods are impacting each of them. It describes the primary mathematical models that are currently available or being developed. On this basis, the committee then assesses the potential impact of HECC on the major challenges of evolutionary biology.

MAJOR CHALLENGES OF EVOLUTIONARY BIOLOGY

Major Challenge 1: Understanding the History of Life

The most fundamental question posed by Major Challenge 1 is this: How did life arise? Despite the large body of scientific literature, this question remains unanswered. Addressing it requires knowledge spanning the physical and biological sciences: chemistry, the Earth sciences, astrophysics, and cellular and molecular biology. A key piece of information would be knowing whether life is indigenous to Earth or exists elsewhere in our solar system. More generally, people in the field ask how the assembly of simple organic compounds led to complex macromolecules and then to self-replicating entities, and what role Earth-bound processes played.

Another unknown is how many species there are on Earth. Systematic biologists have discovered and described about 1.7 million living species. How many more exist in Earth's ecosystems has not been answered satisfactorily, even to within an order of magnitude. Without a better quantification of life's diversity we have only a very incomplete understanding of the distribution of diversity and thus cannot characterize with precision ecosystem structure and function, extinction rates, and the amount of molecular and functional biodiversity. Lack of knowledge about Earth's biodiversity also precludes our potential use of those species and their products.

Ultimately, Major Challenge 1 calls for us to develop an understanding of the tree of life and then to use it. With advances in methods of phylogenetic reconstruction and increasing amounts of new comparative data from DNA sequences, the last decade has seen an unprecedented increase in our knowledge about the phylogenetic relationships of organisms, which collectively constitute the tree of life. Since the beginning of the 1990s, the number of species represented in the gene sequence database GENBank has grown to more than 155,000. If this correlates roughly with the number of species that can be placed on phylogenetic trees using molecular data alone, then the combined number of extinct and living species currently included on phylogenetic trees may approximate 200,000 species. Assuming an increase in the number of researchers and technological advances, it seems safe to expect that between 750,000 and 1 million of Earth's estimated 10 million to 100 million species will be placed on trees within the next decade.

Data sets from individual studies are becoming larger and larger, and many contain information on thousands of species and thousands of molecular and/or morphological characters (qualities or attributes). This situation creates well-known computational challenges when one searches existing phylogenetic trees or works to resolve relationships so as to add or clarify particular branches (Felsenstein, 2004). In addition, as more species are added, more comparative character data must be added in order to resolve relationships with confidence.

Resolving the tree of life has become a priority because the hierarchical pattern of relationships is a powerful predictive tool for comparative biology, with many applications in the fundamental and applied sciences and in industry (Bader et al. 2001; Cracraft, et al. 2002; Cracraft and Donoghue, 2004). Beyond knowing how specific species are related to one another, phylogenetic methods themselves are now routinely incorporated into many fields of evolutionary biology, molecular and developmental biology, the health sciences (comparing viral sequences as well as describing and predicting molecular change), species discovery/description, natural resource management, and biosecurity (identification of invasive species, pathogens). In addition, they are used in the identification of microorganisms, in the development of vaccines, antibacterials, and herbicides, and by the pharmaceutical industry in the prediction of drug targets.

Computational Challenges

Standard phylogenetic analysis comparing the possible evolutionary relationships between two species can be done using the method of maximum parsimony, which assumes that the simplest answer is the best one, or using a model-based approach. The former entails counting character change on alternative phylogenetic trees in order to find the tree that minimizes the number of character transformations. The latter incorporates specific models of character change and uses a minimization criterion to choose among the sampled trees, which often involves finding the tree with the highest likelihood. Counting, or optimizing, change on a tree, whether in a parsimony or model-based framework, is a computationally efficient problem. But sampling all possible trees to find the optimal solution scales precipitously with the number of taxa (or sequences) being analyzed (Felsenstein, 2004) (Figure 4-1).

Thus, it has long been appreciated that finding an exact solution to a phylogenetic problem of even moderate size is NP complete (see, for example, Bader, 2004). Accordingly, numerous algorithms have been introduced to search heuristically across tree space and are widely employed by biological investigators using platforms that range from desktop workstations to supercomputers. These algorithms include methods for fusing and splitting taxa, swapping among branches, and moving through tree space stochastically to avoid becoming stranded at local suboptimal solution sets (Felsenstein, 2004; Yang, 2006). Such methods include simulated annealing, genetic (evolutionary) algorithmic searches, and Bayesian Markov chain Monte Carlo (MCMC) approaches (see, for example, Huelsenbeck et al., 2001), among others.

The accumulation of vast amounts of DNA sequence data and our expanding understanding of molecular evolution have led to the development of increasingly complex models of molecular evolutionary change. As a consequence, the enlarged parameter space required by these molecular models has increased the computational challenges confronting phylogeneticists, particularly in the case of data sets that combine numerous genes, each with their own molecular dynamics.

The growth of phylogenetic research and its empirical database presents computational challenges beyond those of pure tree building. Phylogeneticists are more and more concerned about having statistically sound measures of estimating branch support. In model-based approaches, in particular, such procedures are computationally intensive, and the model structure scales significantly with the size of the number of taxa and the heterogeneity of the data. In addition, more attention is being paid to statistical models of molecular evolution (see, for example, Nielsen, 2005; Yang, 2006), which are the backbone of reconstructing ancestral sequences across a tree. This type of analysis is a prime objective

Species	Number of Trees
1	1
2	1
3	3
4	15
5	105
6	945
7	10,395
8	135,135
9	2,027,025
10	34,459,425
11	654,729,075
12	13,749,310,575
13	316,234,143,225
14	7,905,853,580,625
15	213,458,046,676,875
16	6,190,283,353,629,375
17	191,898,783,962,510,625
18	6,332,659,870,762,850,625
19	221,643,095,476,699,771,875
20	8,200,794,532,637,891,559,375
30	4.9518×10^{38}
40	1.00985×10^{57}
50	2.75292×10^{76}

FIGURE 4-1 The number of rooted, bifurcating, labeled trees for n species, for various values of n. The numbers for more than 20 species are approximate. SOURCE: Felsenstein (2004).

for many evolutionary biologists and has numerous applications, including reconstructing virulence in viruses, predicting the probabilities of genetic change, and the design of vaccines.

Theoretical molecular evolutionists and phylogeneticists have long simulated data sets and trees—to understand and compare phylogenetic methods and their statistical properties, for example, and to compare models of sequence change between simulated phylogenies and those found in the real world. The scale and efficacy of these studies are inherently limited by computational capability as investigators seek to make their simulations more sophisticated and realistic.

Finally, as phylogenetic studies increase in scope, visualization of the results becomes more computationally complex (NSF, 2005b). The problem of visualization has received less attention than other aspects of phylogenetics, but because the field is growing so rapidly, visualization will need to be addressed. The computational challenge associated with visualization calls not only for more computational capability but also for the development of visualization software for phylogenetics, which has received very little attention.

Major Challenge 2: Understanding How Species Originate

The predominant view of how species originate is that speciation takes place through geographic isolation of populations, followed by differentiation, a process known as allopatric speciation. However, an increasing number of biologists are proposing that many species have arisen under local, nonallopatric conditions owing to rapid shifts in host preference, a process known as sympatric speciation, in which the diverging populations are not separated geographically. A third mode of speciation, parapatric, postulates taxonomic differentiation along steep environmental gradients in which divergence can occur under intense natural selection even in the presence of some exchange of genes between species. A fundamental unknown is associated with Major Challenge 2—namely, What is the relative importance of these alternative modes of speciation across different taxa?

The frequency of different modes of speciation across taxonomic groups is critical for proposing and testing general explanations about the genetic, developmental, and demographic conditions leading to speciation, as well as for understanding patterns of species diversity. Even though different modes may have multiple underlying mechanisms, knowledge about mode frequency will form the cornerstone of any general mechanistic theory of speciation. Although the factors leading to allopatry (vicariance[1] and dispersal) are well-known, the conditions under which sympatric and parapatric speciation can occur have received moderate theoretical treatment but not enough experimental and empirical study.

Of course, Major Challenge 2 cannot be clearly defined, let alone addressed, unless we understand the nature of species and their boundaries. Two associated questions are these: What changes take place in the genetic and developmental architectures of isolated populations? What mechanisms underlie those changes? Few questions in evolutionary biology have generated as much debate as how species are to be defined. At a certain fundamental level, these debates are about the nature of species boundaries. One type of boundary involves reproductive relationships within and among populations. Traditionally associated with the notion of biological species, in that case the boundary delimits populations that have evolved genotypic or phenotypic characteristics that make them reproductively incompatible with (isolated from) one another. A second form of boundary, generally associated with phylogenetic or cladistic species, circumscribes populations that are diagnosably distinct from one another, whether or not these character differences result in reproductive isolation.

Leaving aside the issue of which type of boundary should adjudicate the species debate, both types raise fundamental problems of interest to evolutionary biologists. In many taxonomic groups there are multispecies lineages in which viable offspring are produced by hybridization events in the wild, even between species that are distantly related. Although these hybridization events may be uncommon or rare due to pre-mating isolating barriers, they imply that distinct taxa are able to maintain cohesion of their developmental and genetic architectures in the face of some gene flow. Therefore, reproductive incompatibilities are extremely weak or nonexistent, and this situation can be maintained for millions of years. In other groups, by contrast, closely related and very similar species that were isolated only recently and have since become sympatric can have strong reproductive incompatibilities. Thus we are faced with the question of how to explain the continuum of responses to isolation and differentiation in genetic and developmental terms.

Addressing Major Challenge 2 will also necessitate fundamental advances in understanding how changes in genetic architecture translate into changes in the development of the organism and of its phenotypic (observable) characteristics as an adult. It is likely that many types of changes in these complex genetic-developmental pathways could lead to reproductive incompatibilities in behavior,

---

[1] Vicariance describes a situation in which a widespread population is subdivided into two or more allopatric populations by a newly formed physical barrier such as a river, mountain range, or change in habitat due to environmental change.

physiology, or ecological preferences, but we currently do not know to what extent predictive regularities exist. Old debates about the relative importance and frequency of "micro" versus "macro" phenotypic effects of mutational change are, in many respects, still with us because simple genetic changes can potentially be amplified into significant phenotypic differences through complex developmental networks. Evolutionary biologists have long believed that reproductive incompatibilities are more significant in genetic terms than are even striking phenotypic changes that do not result in those incompatibilities. This inference, however, is based more on looking at the world through a particular theoretical framework than on generalizable knowledge about the genetics of differentiation.

The fact that it is possible to maintain the integrity of species even in the face of hybridization indicates those species' genomes are porous but at the same time resistant to gene flow that might break down their boundaries. These considerations raise questions not only about the nature of species and their boundaries but also about the morphogenetic mechanisms responsible for maintaining phenotypic expression. Very little is known about this, and answers will require an approach calling on the expertise of numerous disciplines.

A third key component of Major Challenge 2 is learning about the history of populations and species, particularly the human species. As a consequence of the vast increase in genomic information, these histories can be probed in increasing detail. But populations and species are complex entities; every individual of every species has thousands of genes in its genome. Combining genomic data therefore becomes an issue: ribosomal DNA sequences from organelles like chloroplasts and mitochondria may yield different histories from those of the nuclear genome. These genes have variable mutation rates and can provide varying amounts of information for demographic history. Whereas traditional phylogenetic analysis often treats each species as a single monomorphic entity that cannot be further decomposed, a genomic perspective of biodiversity acknowledges that each species is in fact a complex entity whose cohesion and trajectory through time could be constrained by many processes, including reproductive isolation from other species. The theoretical framework provided by coalescent theory—a body of population genetics theory that models the genealogical signatures of genetic lineages as they are passed down through generations—has further increased the resolving power of DNA sequence data.

Computational Challenges

Much current research on the above questions and problems can and does proceed without extensive computational requirements. The relative frequency of modes of speciation, for example, can often be determined by standard phylogenetic and biogeographic studies. And there have been numerous computational studies in population biology and genetics that attempt to model the conditions under which sympatric speciation is likely to operate (reviewed in Gavrilets [2003] and in Coyne and Orr [2004]).

As evolutionary biology in general and speciation research in particular become more mature and as new genomic information becomes available, there has been a shift to the use of mathematical models and methods. Currently, there is growing interest in developing mathematical models for particular cases in order to test well-defined hypotheses associated with speciation—for example, was speciation of cichlids in a Nicaraguan crater lake sympatric, or was it a result of double invasion? Did modern humans hybridize with Neanderthals during their colonization of Europe? While many of the current studies do not require high-end computing, future advances in population evolution and speciation will be stymied unless they can scale, which will require new computational approaches, algorithmically and computationally.

Coalescent models now encompass a wide variety of demographic and genetic phenomena, including population bottlenecks, migration, change in population size over time, natural selection, gene flow between populations, gene conversion and recombination between alleles in a population, and complex mutation patterns. DNA sequence data are now routinely analyzed in terms of genealogical (phylogenetic) trees at various levels in the hierarchy, from individuals within a population to closely and distantly related species. These genealogical patterns will show signatures of various demographic processes, including migration, population size changes, and reproductive isolation events through time. However, the highly stochastic nature of the coalescent process means that the realization of any particular genealogical pattern in nature could be consistent with many different scenarios. As a result, genealogical signatures from many different genetic loci are required to accurately estimate demographic histories. Linking many different genealogical signals with models of various population demographics is computationally demanding, and advances to date have been made via a series of strong but useful approximations. Relaxing these assumptions and exploring the full range of demographic histories is a major computational challenge for the future (Beaumont, 2002, 2004).

Going hand in hand with these developments, however, is a pressing need for gathering more empirical data, which can provide the context for building more realistic models. These types of data would include the kinds and amount of genetic variation and selective pressures that might bear on the evolution of pre- and postzygotic reproductive isolation.

It seems clear that studies along the interface of population biology and efforts to unravel the origin of species will make significant use of computational resources. Increased understanding of how populations become spatially structured genetically will rely on large populational sampling and detailed descriptions of populational history. Moreover, the integration of genetic and demographic information through complex models and simulations of populational histories will present profound computational challenges.

Major Challenge 3: Understanding Diversification of Life Across Space and Time

At a general level, it is well known that processes in the geosphere and biosphere have been tightly linked since the origin of life (NRC, 1995), but we have only partial understanding of the linkages across different spatial and temporal scales. At large scales, movements of continents and terrains, tectonic effects within continents, and long-term climate changes have had a profound influence on the distributions of organisms and the ecological associations they comprise, and such phenomena may be first-order drivers of biotic evolution. At smaller geographic and temporal scales, geological processes can be implicated in controlling the rates of speciation and extinction. At still smaller scales, geospheric processes influence local environmental change, which is one cause of microevolutionary change within and among populations. At none of these scales do we have enough theory on which to build strong models or the ability to simulate the coupled processes.

Beyond understanding the coupling, we would like to understand the intrinsic and extrinsic controls on the rate of speciation. Over long periods of time, diversity has increased, decreased, or remained relatively stable, yet at a mechanistic level the causes of these patterns are poorly known. Diversification is generally modeled as a birth/death process, with change in diversity over time being a function of the rate of speciation and extinction. Many factors have been implicated in the rate controls of each, but current models describe simple diversity-dependent processes that rely largely on biotic interactions as proximate causes of increases or decreases in speciation and extinction. Omitted are abiotic factors such as tectonically mediated changes in mountain building, large-scale alterations of river systems, or climate change, all of which are widely recognized as potential drivers of rate controls.

There is a pressing need for more realistic models of diversification that can be applicable across different spatiotemporal scales. These might not only parameterize traditional biotic rate controls but also take into account causal linkages between Earth history and speciation/extinction rates. This also provides a foundation for understanding how communities and ecosystems are assembled across space and time.

Understanding the evolutionary assembly of communities and ecosystems is a fundamental problem that cuts across multiple disciplines, including systematics and historical biogeography, community and landscape ecology, and paleontology. It has application to conservation, resource management, and understanding the consequences of global change. The mechanisms governing the assembly and maintenance of species associations (communities, ecosystems) at different spatial scales have received considerable attention, especially in ecological science (see, for example, Ricklefs and Schluter, 1993), but many aspects of the evolutionary dynamics of these assemblages have been less studied. Increasingly, history is recognized as playing an important role in shaping taxonomic assembly at a wide range of scales. The coevolutionary history of different groups, or clades, of organisms within biological communities can be analyzed using methods of historical biogeography, but there is considerable disagreement and little consensus on whether any of the current methods are sufficiently sophisticated to reconstruct the spatial history of moderate to large clades, let alone multiple clades simultaneously. Developing a more sophisticated understanding of community assembly over multiple timescales will necessitate the development of new models and algorithms to integrate multiple species histories and ecologies.

A related part of Major Challenge 3 is how to determine the evolutionary history of microorganismal community structure and function. This is a somewhat different challenge, because new methods in comparative genomics are giving us a better understanding of microbial community organization. Termed "environmental genomics" or metagenomics, these new tools use advances in high-throughput sequencing to sample the genomes in environmental samples (Riesenfeld et al., 2004; NRC, 2007). Conventionally, microbial DNA is isolated from a sample, cloned, and then used to create metagenomic libraries, but newer technologies can access the environmental sample directly (NRC, 2007). The resulting sequences have many uses, including for phylogenetic studies, measures of taxonomic and genomic diversity, the discovery of new genes, functional analysis of specific genes, and for modeling large biochemical pathways, including community metabolism (Tyson et al., 2004; Rusch et al., 2007).

Although current metagenomics research is primarily directed toward genome characterization and the structure and function of microbial communities, it is clear that the massive amounts of new sequence data being collected will substantially expand our knowledge of global diversity, the tree of life, and genome evolution. Additionally, genomic comparisons interpreted in the context of phylogenetic relationships can also be expected to reveal new insights into genome structure and function.

Computational Challenges

Many of the causal linkages between the geosphere and biosphere, such as how tectonically driven change might have influenced the speciation and extinction rates, have been inferred from correlations generated by empirical studies. There is a need for a better theoretical-mathematical foundation that can lead to predictive quantitative assessment. Moreover, these causal models linking Earth history with biotic evolution should be operable at different spatiotemporal scales. Some simulations have been run that couple climate models with environmental models and data about ecosystems, and modeling of species distribution is becoming increasingly common. Some simulations have been performed to reconstruct how taxonomic elements of communities and ecosystems have assembled over time by integrating the phylogenetic and spatial histories of many groups of organisms simultaneously. To date, however, few if any of these studies have taken advantage of high-capacity computing.

Major Challenge 4: Understanding the Origin and Evolution of the Phenotype

The phenotype describes the observable characteristics of an individual. Although an individual may have the genetic capability to express a trait, only the traits that manifest are considered phenotypic characters. Phenotypic characters can be broadly construed across different levels of organization, from genomic and developmental characteristics to external features, physiology, and behavior. The genetic architecture of individuals or populations includes all interactions and functional linkages among genes that have influence on the expression of traits (NSF, 1998). We have broad knowledge of the nature of genetic variation within and between populations, and the promise of large-scale genomic sequencing across many individuals promises to revolutionize that knowledge and allow much more sophisticated questions to be asked. Knowledge about the origin and evolution of phenotypes is built on an understanding of genetic variation in all its components, from nucleotide polymorphisms and their frequency in populations to the interactions among genomic loci. A key question is, How does that variation in coding or regulatory genes relate to changes in phenotype?

Recently, significant progress in understanding the evolution of phenotypes has come from integrating the fields of evolutionary and developmental biology. Both fields have long histories, but starting several decades ago they diverged. Evolutionary biologists focused increasingly on understanding evolution at the population level and developed sophisticated genetic models to understand changes in allele frequencies, while developmental biologists focused on experimental manipulations to uncover the mechanisms of development. More recently, however, developmental biologists have taken their analysis to very deep molecular and genetic levels, and this has led to a renewed interest in understanding the interplay between evolution and development (called "evo-devo"), as both fields now have a common language of genetics and genomics. Evo-devo has recently had explosive growth and has become an exciting area of investigation and attracted much popular attention, from, among others, S.B. Carroll (2005).

These studies of developmental evolution have spanned all levels, from microevolution within populations to macroevolution among the major clades of life. One of the remarkable outcomes of these initial studies is the discovery that individual genes and genetic pathways can have important evolutionary effects on development and morphology. For example, patterning along the anterior-posterior axis of animals is controlled by a set of genes known as the homeotic (*Hox*) genes. While initially characterized through the study of highly deleterious mutant alleles in model species such as *Drosophila melanogaster* (fruit fly), subsequent studies have shown that evolutionary changes in the *Hox* genes and changes in how those genes are expressed play a clear role in animal evolution across the entire micro- to macro-evolutionary spectrum. For example, the *Hox* gene *Ultrabithorax* (*Ubx*) has helped us to understand microevolutionary changes in the pattern of fly bristles as well as the macroevolutionary changes seen in the appendages of crustaceans (Stern, 1998; Averof and Patel, 1997). Similarly, *Hox* genes have also been implicated in the evolution of large-scale changes in the vertebrate skeleton during evolution. The analysis of these genetic networks presents us with the opportunity to understand evolution in increasingly sophisticated ways and has allowed us to generate better models of how evolutionary change occurs.

For quite some time, evolutionary models suggested that the phenotypic differences between even very closely related species were due to variation at a large number of genomic loci, with any given individual mutation having only a minute effect. Recent experiments, however, suggest that this is not always the case. Increasingly we see that phenotypic variation can often be attributed to one or just a few genes of large effect. For example, recent studies in sticklebacks show that variation at a single gene, *Pitx1*, has a very large effect on the pelvic spines found in these fish (Shapiro et al., 2004). At the molecular level, these changes can involve both protein coding and gene regulatory modifications, although current theories suggest that regulatory mutations in developmental genes may have a predominant role in underlying major evolutionary change in phenotype. The pace of such discoveries is ever increasing as new genomic and developmental tools are allowing us to decipher how developmental systems evolve.

One fundamental goal of Major Challenge 4 is to understand integrated phenotypes and how they evolve. Phenotypic features—such as morphological form, physiology, behavior, even biochemical pathways—are often integrated into functional groups based on their interaction with the environment. Such linkages may be tight or loose. An additional challenge is to understand the boundaries and strength of these linkages, how the components of these groups arise, coevolve, and perhaps become unlinked functionally, potentially to be incorporated into other functional groups. Thus, considerable attention is now being paid to establishing the boundaries of integrated suites of morphological, behavioral, and physiological traits of organisms that function and interact collectively with the environment. The concept of modularity within evolutionary developmental biology is playing a role in understanding the structural and functional organization of integrated phenotypes and how they arise in development and change during evolution (Schlosser and Wagner, 2004). Integrated phenotypes are being studied from the perspectives of developmental biologists, comparative functional biologists, and population biologists conducting ecological and genetic experiments in the wild or laboratory.

These evolutionary alterations in developmentally important genes also lead to changes in the genetic architecture of development in such a way as to control the range of phenotypic variation that is possible in subsequent generations. In some situations this can constrain or limit future phenotypic evolution, while in other cases it can open up entirely new possibilities for subsequent evolutionary change and the appearance of totally novel morphologies and physiologies. An important future challenge is to integrate these developmental and evolutionary studies with ecological ones to understand how natural selection shapes the course of growth and development of phenotypes and the underlying genetic architecture.

Computational Challenges

A serious major computational challenge is to generate qualitative and quantitative models of development as a necessary prelude to applying sophisticated evolutionary models to understand how developmental processes evolve. Developmental biologists are just beginning to create the algorithms they need for such analyses, based on relatively simple reaction-rate equations, but progress is rapid, and this work will soon be able to take advantage of HECC resources.

Another important breakthrough in the field is the analysis of gene regulatory networks (Levine and Davidson, 2005). These networks describe the pathways and interactions that guide development, and while their formulation is dependent on intense experimental data collection, once produced, the networks provide explicit models to test how perturbations affect all manner of developmental events. While they resemble metabolic pathways in overall structure, they are far more complex in their regulation and behavior. As these models grow to include more pathways and more organisms, they will increasingly benefit from greater computational capacity and will become vital to many evolutionary studies. Similarly, protein-interaction network analysis provides insight into an organism's functional organization and evolutionary behavior—see, for example, http://www.hicomb.org/papers/HICOMB2007-03.pdf.

Major Challenge 5: Understanding the Evolutionary Dynamics of the Phenotype–Environment Interface

Evolutionary biology has long sought to understand environmental selective effects by focusing on a single trait or trait complex, such as bill shape, body size, or body shape. Yet, selective regimes in the environment act on entire phenotypes that are the result of highly complex and linked (integrated) developmental and metabolic pathways. Earth's biota is a product of complex interactions between the abiotic and biotic realms (see Major Challenge 3). Life on Earth has evolved in the context of a dramatically and often rapidly changing environment, whose trajectory and long-term trends have themselves been modified by evolving life forms. Earth's atmosphere and long-term climate trends have played an important role in the major transitions and increasing complexity of life, but the system is interactive, replete with complex positive and negative feedbacks between living and geological systems. The timing and causes of many of the major transitions in the origin of biotic complexities, such as the origin of oxygen-based metabolism, are still somewhat controversial, given the challenges of interpreting the chemical and morphological signals in the fossil record of the first 3 billion years of evolution.

The evolution of integrated complexes can also be investigated at different hierarchical levels. One critical approach is to link variation in these integrated complexes to environmental differences within and among populations in order to understand outcomes of selection. At a higher level, changes in integrated complexes can be analyzed across species, particularly those that are closely related, thus describing how the components of these complexes change at times of speciation. Such analyses are critical for providing insight into how the tightness of integration "constrains" change in phenotypic-functional complexes.

Finally, we build on this knowledge to learn about the relationship between phenotypic change and adaptation. At a population level, variants in phenotype may have different consequences for survival or reproduction. Those variants that become fixed through natural selection (because of those consequences) are often referred to as adaptations. The nature of adaptations has been studied intensely from the viewpoint of population biology and ecology. Less attention has been paid to the molecular basis of population variation underlying phenotypic change and the linkages that might exist between genome evolution and phenotypic evolution. To what extent, for example, is convergence in phenotype related to convergence in the genetic and developmental pathways that produce those phenotypes? And, to what degree is similarity in presumptive adaptations constrained by those pathways, or is there flexibility in morphogenetic systems such that different ones can produce very similar phenotypic expressions? Answers to many of these questions and others in this field will require more empirical information about the amount and kind of genetic variation that underlies phenotypic variation and its response to selection. Such data are crucial for building more sophisticated and realistic models and simulations of the evolutionary process.

A second fundamental goal underlying Major Challenge 5 is to understand better the links between ecological and evolutionary processes. The conservation of biodiversity depends critically on our ability to predict the responses of populations to changes in their environment that occur on short- and medium-term timescales. It is likely that models can be developed that are capable of predicting short- and medium-term population fluctuations in response to environmental change in greater detail than we can over long-term evolutionary time. Such models could, for instance, address biologists' concerns about the effects of climate change in the recent past and in the short-term future. Both the increased detail of the environmental record and the increased sophistication of demographic models should enable biologists to understand the effect of environmental change on many of the key life-history components of population fluctuations, such as juvenile and adult survival, fecundity, and population age-structure.

From their humble beginnings as simple logistic growth curves, ecological models now routinely grapple with the effects of stochastic environmental change on demographic and population trends. But populations do not expand indefinitely in favorable environments; they inevitably hit the carrying capacity of the ecosystem and undergo declines due to increased competition. Many such density-dependent population crashes have been documented in real populations that have been the subject of long-term censuses. Thus, most models now acknowledge the ecological feedbacks intrinsic to the dynamics of populations, in particular the phenomenon of density dependence. Other demographic complexities are also being incorporated into ecological models, such as age-specific responses to the environment, variable levels of year-to-year autocorrelation in environmental variables, and cohort responses, in which the environmental conditions specific to the birth year of juveniles affects the shapes of their survival and fecundity curves (Lindstrom and Kokko, 2002). All of these variables impose additional complexities for predicting population responses to environmental change. Thus realistic models of population and demographic history must take into account stochastic changes in the environment, as well as the impacts of a population's growth on the environment and associated feedbacks. For example, a recent simulation study (Wilmers et al., 2007) found that drastic fluctuations in populations, and hence increased chances of extinction, were more likely to be found in environments that were positively correlated from year to year.

Computational Challenges

Increasingly, evolutionary biologists are incorporating realistic models of population and demographic changes into their modeling of population genetic processes (Whitlock and Gomulkiewicz, 2005). In many respects, ecological and evolutionary modeling have proceeded independently of one another, primarily because of the mathematical difficulties of linking the two holistically. One difficulty is that ecological and evolutionary processes are perceived to operate on vastly different timescales. In addition, the dynamical complexity of population responses to environmental change observed in many ecological models typically exceeds the complexity of the types of demographic changes modeled in population genetics. For example, relatively few population-genetic models cover situations in which populations are not at equilibrium. Even complex migration models assume long-term stability of the migration matrix and population sizes over time. Although models that attempt to estimate population-size changes and rates of growth and decline from molecular data do address a type of nonequilibrium situation, they are limited to estimating, for example, a constant rate of population growth. Typically, parameters of interest are estimated using computationally burdensome methods such as maximum likelihood, in which the optimal parameter values are searched for heuristically to optimize the likelihood. Alternatively, a faster approach involves sampling multiple parameter values stochastically, evaluating the likelihood of each, and accepting these in proportion to their likelihood or their posterior probability. This is the essence of Bayesian approaches. Evaluating the likelihood of a set of parameter values is not in itself time consuming, but finding the optimal set of values is.

Recent approaches to inferring past population dynamics from multiple genetic loci now use simulation of gene genealogies within a specified population model and compare simulated patterns of DNA sequence diversity to the patterns observed in the data under study. Such simulations typically utilize a relatively simple population model. The fit of these simulated DNA sequences to the data actually collected can be assessed using summary statistics, and the efficiency of simulations can be increased through statistical methods such as importance sampling. Typically, this involves first simulating gene trees within population histories and then simulating mutations on these trees in order to produce DNA sequences consistent with the model. Incorporating all the demographic events that have been proposed to influence populations is computationally challenging.

Lewontin (1979) expressed the dynamic interplay between ecological and evolutionary and population genetic processes through a set of interrelated differential equations:

$$d\mathbf{N}/dt = f(\mathbf{N}, P) \text{ and } d\mathbf{G}/dt = g(\mathbf{G}, \mathbf{W}).$$

Here the vector $\mathbf{N}$ represents the distribution and abundances of species in nature, $\mathbf{G}$ represents the genetic structures of those same species, and $\mathbf{W}$ represents the parameters of selection governing the microevolutionary trajectories of the species. In the first equation, P represents the parameters of population growth and interactions among species as mediated by morphology, physiology, dispersal, and so on. The key insight is that the parameters making up P are in turn a function of $\mathbf{G}$; that is, $P = h(\mathbf{G})$. In a similar fashion, microevolution ($\mathbf{W}$) is determined in part by the abundances of species, $\mathbf{N}$: $\mathbf{W} = H(\mathbf{N})$. That is, evolution is density dependent.

These relationships between P and $\mathbf{G}$ and between $\mathbf{W}$ and $\mathbf{N}$ express the direct coupling of ecological and evolutionary processes. The integration of ecological and evolutionary modeling is still incomplete, though. As noted above, ecological theory ($d\mathbf{N}/dt$) and evolutionary theory ($d\mathbf{G}/dt$) are typically studied in isolation from each other today; thus one of the main objectives of computational research at the interface of ecology and evolution will be the "recoupling" of ecological and evolutionary dynamics. Such approaches have contributed to some solid conceptual advances, and they hint at how the reintegration of ecology and evolution can lead to results that would not be expected had they been studied in isolation. One such unexpected result is the prediction of mutational meltdown (extinction due to accumulation of deleterious mutations) in metapopulations in situations when population sizes become critically low, thereby accelerating the fixation of deleterious mutations.

Major Challenge 6: Understanding the Patterns and Mechanisms of Genome Evolution

No tool kit has revolutionized evolutionary biology more than genomics. The foundation of large-scale genome analysis is the complete sequencing of genomes, whether from single-celled bacteria or unicellular eukaryotes or from more structurally and developmentally complex animals and plants. Such an approach allows systems-level comparisons of whole genomes across evolutionary time and across lineages that share few or no morphological traits. The advent of ever-more-rapid sequencing approaches promises that our ability to compare whole genomes to one another will only improve and provide an increasingly refined picture of the processes underlying genome evolution at all scales.

Yet very fundamental questions remain, such as understanding how new genes arise. There are thought to be a large number of often serendipitous routes by which new genes can arise (Lynch, 2007). Duplication of individual genes, entire chromosomes, or entire genomes, followed by subsequent divergence, is probably the most common mechanism for the generation of novel genes. As a consequence of this mechanism, the new genes are usually closely related in sequence to their progenitor genes. The fate of these genes will depend on the selective environment at the time of duplication and the availability of favorable mutations. Newly created genes often experience a host of novel mutations: the expansion and contraction of repeated amino acid motifs; changes to gene regulatory domains, which control when and where the genes are expressed; additions and deletions of entire exons and domains; and even changes in reading frame.

The role of whole-genome duplications is also an important consideration in organismal evolution, especially in vertebrates and certain plant lineages. The fate of genes after duplication, either as part of whole genome duplications or just individual gene duplications, has been extensively modeled, but as we get more and more genomic data we will begin to get a better picture of the actual course of events and the role of genetic change in evolutionary change. Several large-scale surveys of gene duplications appeared in recent years, and the estimated rates of gene duplications are now thought to be about the same as, if not faster than, rates of point mutations in DNA sequences (Lynch, 2007).

Transposable elements are a diverse class of autonomous mobile DNA elements that can proliferate within genomes, copying themselves from sites of origin to novel chromosomal positions at bewilderingly high rates. Transposable elements have been known since the mid-1900s, yet their full impact on the evolution of genomes and organisms has only recently been appreciated. Approximately 40 percent of the human genome is composed of easily recognized transposable elements, and if one includes ancient elements whose identity has eroded over time, the fraction affected is likely much greater. As such, they are probably the single most important long-term component of complex eukaryotic genomes, and their effects on evolution and disease are profound and widespread (Batzer and Deininger, 2002; Deininger and Batzer, 2002). Transposable elements have long been known as a source of mutation, and many of their early phenotypic and genetic effects were considered detrimental or at best neutral. However, in the last 5 years, the number of cases in which transposable elements have played a demonstrably adaptive role in the origin of new genes and genomic functions has dramatically increased, so that traces of transposable elements in the genome can no longer be assumed to be junk DNA.

Transposition, often through an mRNA intermediate, is responsible for the many genes in the genome without introns, so-called retrotransposed genes (Brosius, 1999). These novel genes often assume novel functions unrelated to the function of the original gene. As a result, the creation of novel genes via transposition is thought to underlie some of the most dramatic examples of adaptation to extreme environments, such as the ability of notothenioid ice fish to maintain their blood in liquid form, without freezing, while inhabiting the subfreezing Antarctic Ocean (Chen et al., 1997).

Transposable elements are now known to provide some of the basis for novel *cis*-regulatory elements of genes, enabling enhanced versatility at the level of expression. The structures of many genes clearly retain vestiges of transposable elements, indicating that the genome frequently co-opts these elements for novel coding functions. Sometimes the breaking up and reregulation of genes by transposable elements clearly has adaptive value; it has been postulated that the vertebrate immune system was brought about by the insertion of a transposable element into an ancestral immunity gene. At the same time, it has been estimated that up to 10 percent of congenital diseases in mice are caused by transposable elements. The balance between adaptive and detrimental consequences for transposable element proliferation is thus a key unknown in evolutionary genomics.

Transposable elements are also an important factor in the variability of total genome size of animals, plants, and fungi. For example, the evolution of the vertebrate genome over the past 600 million years has seen the origin, proliferation, and decline of several families of transposable elements. Why families of transposable elements increase and then decline, often in concert with changes in the numerical dominance of unrelated families, is still unclear. Are there global genomic regulations on the number of transposable elements that are driven by the deleterious effects of accumulation? If so, where do these regulations come from and can they be predicted? What is the ecology of transposable elements in the genomic community? Large-scale variation in transposable elements will play an important role in explaining the 1,000-fold variation in genome size observed among living eukaryotes.

As we build our knowledge of how genes arise, evolutionary biology can begin to understand how evolution is constrained by networks of interaction between genes and noncoding elements in the genome. Genes do not exist in isolation; they belong to complex networks of interacting genes, gene products, and environmental signals. There is increasing interest in the degree to which genetic networks are robust to perturbations from within and without. Perturbations from within include the generation of novel genes as a result of gene duplication, and the fate of such gene duplicates is known to depend on the novelty of function of the new gene and the structure of the network into which it is inserted.

Perturbations from without include environmental changes, and the effect of such perturbations on the stability of developmental modes and outcomes is key to understanding whether networks are adaptive or whether they have arbitrary and neutral aspects to their construction. Genetic networks are also known to influence rates of genomic change (Barton et al., 2007), and some possess a structure that buffers against the insertion of new genes and gene functions.

In recent years biologists have added a host of novel interacting genomic elements to the already long list of well-annotated genes. Such novelties include "ultraconserved" noncoding elements in the human genome, which are short sequences—fewer than 200 base pairs (bp)—that are 100 percent conserved between humans, other mammals, and in some cases chicken and fish; microRNAs, which are short sequences encoding RNAs of only ~22 bp that are now known to play critical roles in development and gene regulation (Bartel, 2004); and long-distance enhancers, short regions of the genome that regulate a set of constituent genes despite sometimes being located millions of base pairs away from their targets. Untangling the web of interactions among such diverse sets of genomic players and providing a seamless link between genomic data (such as large-scale gene expression data) and network models remains a computational challenge for the future.

Progress over the last 10 years has refined to an unprecedented degree our characterization of primary genome constituents in humans and other organisms while revealing new structures and forces operating within complex genomes that were unknown in the pregenomic era. Understanding how these myriad constituents interact and influence one another and how genomes and chromosomes function and evolve as hierarchical networks is a major challenge for evolutionary biology. Many of the principles of population genetics and molecular evolution, laid down in the twentieth century, are still applicable to genome data despite the scaling up from single genes to entire genomes (Li, 1997; Lynch, 2003). By contrast, the computation and estimation of whole-genome parameters and the analysis of data sets that are vastly increased in size has required, and will continue to require, new computational and algorithmic tools.

Computational Challenges

Obtaining genomic sequences is becoming easier and less expensive each day, and we are becoming inundated with genomic data. It is predicted that within a few years the cost of sequencing a human genome will drop to less than $1,000 and that such information will be a routine component of personal medical information used for diagnostic purposes. Making sense of all these sequence data, however, requires a combination of computational and evolutionary approaches.

Modern sequencing technologies routinely yield relatively short fragments of a genomic sequence, from 25 to 1,000 bp. Whole genomes range in size from the typical microbial sequence, which has millions of base pairs, to plant and animal sequences, which often consist of billions of base pairs. Additionally, "metagenomic" sequencing from environmental samples often mixes fragments from dozens to hundreds of different species and/or ecotypes. The challenge is to take these short subsequences and assemble them to reconstruct the genomes of species and/or ecosystems (NRC, 2007). While the fragment assembly problem is NP-complete, heuristic algorithms have produced high-quality reconstructions of hundreds of genomes. The recent trend is toward methods of sequencing that can inexpensively generate large numbers (hundreds of millions) of ultrashort sequences (25-50 bp). Technical and algorithmic challenges include the following:

- Parallelization of all-against-all fragment alignment computations.
- Development of methods to traverse the resulting graphs of fragment alignments to maximize some feature of the assembly path.

- Heuristic pruning of the fragment alignment graph to eliminate experimentally inconsistent subpaths.
- Signal processing of raw sequencing data to produce higher quality fragment sequences and better characterization of their error profiles.
- Development of new representations of the sequence-assembly problem—for example, string graphs that represent data and assembly in terms of words within the dataset.
- Alignment of error-prone resequencing data from a population of individuals against a reference genome to identify and characterize individual variations in the face of noisy data.
- Demonstration that the new methodologies are feasible, by producing and analyzing suites of simulated data sets.

Once we have a reconstructed genomic or metagenomic sequence, a further challenge is to identify and characterize its functional elements: protein-coding genes; noncoding genes, including a variety of small RNAs; and regulatory elements that control gene expression, splicing, and chromatin structure. Algorithms to identify these functional regions use both statistical signals intrinsic to the sequence that are characteristic of a particular type of functional region and comparative analyses of closely and/or distantly related sequences. Signal-detection methods have focused on hidden Markov models and variations on them. Secondary structure calculations take advantage of stochastic, context-free grammars to represent long-range structural correlations.

Comparative methods require the development of efficient alignment methods and sophisticated statistical models for sequence evolution that are often intended to quantitatively model the likelihood of a detected alignment given a specific model of evolution. While earlier models treated each position independently, as large data sets became available the trend is now to incorporate correlations between sites. To compare dozens of related sequences, phylogenetic methods must be integrated with signal detection.

Major Challenge 7: Understanding the Evolutionary Dynamics of Coevolving Systems

Individuals of the same species or of different species generally have either conflicting or cooperating (mutualistic) interactions. Increasingly, many of these interactions have been found to evolve in relation to one another, a process known as coevolution. These coevolving interactions take many forms, from the symbiosis of organelles that once invaded free-living microbes, to predator-prey, host-parasite, and plant-pollination systems, to cooperative breeders or sexually selected mate choice, and to competitive interactors within habitats and communities. Understanding the evolutionary biology of such interactions has broad implications for solving problems in many areas of applied biology, including human health, agriculture, and resource management. For this reason, there is great interest in understanding the genomic underpinnings of conflict and cooperation and how conflict and cooperation evolve.

Genetic mechanisms underlying conflict and cooperation have long been investigated empirically and theoretically, and considerable research has been undertaken on the genetics of behavior. It is widely recognized that behavior is influenced in complicated ways by numerous genes and their interactions, but we still have inadequate empirical knowledge of the genetic variation in the wild that is available to selection. The new tools of genomics promise to broaden the kinds of questions and approaches that have been standard in the field. To put the interconnections among genomic data, development, neurological function, and expressed behavior in an evolutionary context, studies will need to be comparative. These studies will ask new questions about the numbers and kinds of genes and about differences in the genetic architectures that influence conflictual and cooperative behaviors, including the genetic basis of instinct.

Cross-species, coevolutionary comparisons of multispecies interactors will reveal new insights into the nature of coevolution itself—for example, How fast is genetic and phenotypic change in interactors? Are coevolving systems conservative over time? How are those systems shaped by genetic factors?

The evolution of conflict and cooperation can be studied at different hierarchical levels. For example, the history of species associations, such as hosts and their parasites, has been the focus of considerable phylogenetic coevolutionary analysis. At the level of populations, ecologists and population biologists have also built a large library of detailed field and laboratory studies.

Despite this large body of work on conflict, cooperation, and coevolution, many aspects of the roles that conflict and cooperation may play in evolution are still poorly understood—for example, in the evolution of adaptation and in the origin of species. And there is a need for studies that integrate causality at the genomic level with that at the population, ecological, and demographic levels.

Computational Challenges

Sophisticated modeling of conflict/cooperation systems has a long history, and the large body of literature investigating these systems integrates genetic and population approaches. Much of this modeling stems from game-theoretic approaches and from classic population dynamic models such as predator-prey. Although game theory originally dealt with economic problems, it has also had a profound impact on evolutionary biology (see, for example, Maynard-Smith, 1982; Vincent and Brown, 2005). Most quantitative analyses of conflict/cooperation models have been carried out using desktop computing. Yet, as models subsume more parameters and include demographic or genetic information across space and multiple generations, access to advanced capability computing will become necessary (see, for instance, Nowak, 2006).

MAJOR CHALLENGES IN EVOLUTIONARY BIOLOGY THAT REQUIRE HECC

Progress in most areas of evolutionary biology has been very rapid over the past several decades. However, because desktop computing has continued to advance, much of quantitative and theoretical evolutionary biology has prospered without reliance on advanced computational capabilities. But this is likely to be a transient phase, because over the past decade, evolutionary biology has become increasingly multidisciplinary and integrative. This, and the rapid accumulation of genetic and other data on populations and species, has accelerated the transformation of evolutionary biology into a quantitative science. The study of microevolution, which requires genetic and demographic analyses of evolutionary change within populations, has a long history of theoretical and quantitative modeling and remains robust. In many other areas of evolutionary biology, however, the building of quantitative theoretical models has been neglected, and research is sorely needed.

One area that has made use of high-performance machines is phylogenetic research. As more members of the community become adept at using cutting-edge computational methods, there will inevitably be pressure to port models and codes to more powerful platforms and thereby address scientific questions in ways that mirror the complexity of natural systems. Evolutionary biology is already in transition, and to realize its potential fully will require progress on all capability-computing-related fronts: models, theory, data management, education and training, algorithms, and hardware.

It should be stressed that progress in some areas of evolutionary biology is being limited by a lack of computational power. Even in phylogenetic research, where HECC is being exploited, researchers could make use of additional computational resources for increased statistical testing, larger simulations, and advanced visualization tools.

Major Challenges 1 and 2 are probably where we will first feel the need to transition to HECC-enabled research. In both cases, access to advanced computational approaches will enable representing enough complexity to reveal new phenomena. To address Major Challenge 1, the phylogenetics community is building larger and larger trees, evaluating them statistically, and manipulating them visually. As was noted earlier in this chapter, however, the computational challenge scales superexponentially with the number of terminals on the trees. This makes it especially important for the community to take maximum advantage of whatever state-of-the-art computing exists at the time. Not doing so will seriously hamper future progress. In addition, research using phylogenetic methodologies is expanding rapidly in the biomedical sciences as well as for metagenomic studies investigating community structure/function, ecosystem metabolism, and global climate change. In all of these areas, producing results that can meaningfully answer practical questions calls for much greater complexity, which in turn demands advanced computing. Along with the greater complexity, these applications typically involve massive amounts of data, the management and analysis of which will require advanced computing.

For studies about speciation (Major Challenge 2), the simple mathematical models with only a few parameters that were of such importance for previous theoretical work are unable to make good use of the large amounts of data becoming available. They also are proving inadequate given the desire for higher resolution descriptions of genetic and/or population behavior. Thus, although experimentation may continue to be the dominant research modality for Major Challenge 2, there is a need to move to simulations that are explicitly genetic and characterized by a large number of parameters, large populations (hundreds of thousands of individuals), long timescales (hundreds of thousands of generations), and significant stochasticity (which requires that the simulations be run multiple times to enable statistical analysis). For the research community to take the step from qualitative predictions and speculations to much more powerful and precise quantitative predictions and estimates, it must be able to perform such simulations, for which it will need HECC.

Moreover, advanced computing opens some new options for approaching Major Challenge 2. Evaluating demographic histories using genetic data is computationally challenging for several reasons. For any data set of DNA sequences (or other genetic markers), there are many potential gene trees (phylogenies) that are consistent with the data; in addition, for any given set of phylogenies, there are a number of often complex population histories that can be accommodated by this set of gene trees. The result is two levels of uncertainty. In theory, this challenge could be met by integrating across all possible gene trees and population histories:

$$\Pr(X|\Theta) = \int_{G \in \psi} \Pr(X|G) p(G|\Theta)$$

(Felsenstein, 1988; Hey and Nielsen, 2007). Here, X is the collected data (say, DNA sequence data sampled from multiple loci and multiple species or populations); Θ is the species history, which could be a complex demographic history involving bottlenecks, gene flow, and local extinction or a purely dichotomous history of population divergence, a phylogeny; G is a gene tree, an estimated genealogy of DNA sequences; and ψ is the set of all such genealogies, which include continuous branch lengths and very many topologies.

However, this integration is not realizable in practice, not only because of the need for efficient estimation when the state space is large (for complex demographic models with many parameters) but also because, as discussed in connection with Major Challenge 1, the number of possible gene trees (phylogenies) increases superexponentially with the increasing number of taxa (in our case, genes or alleles sampled from a given species). This makes it impractical to evaluate the above integral by sampling a large number of genealogies at random. Thus even though the probabilities of mutation events for any one genetic locus can be multiplied by those for other loci because they are independent (conditional on the demographic history itself), they cannot be calculated analytically.

As a result, computational and statistical approximations have been used extensively. The most recent of these methods to be employed, Bayesian analysis, frequently utilizes MCMC methods to sample many possible gene trees and parameters associated with the demographic history. Various sampling and rejection schemes have been proposed, as well as means of proposing parameters via complex prior distributions. With a sufficient number of MCMC cycles, complex probability distributions can be approximated. Still, this exercise gets us only to the point where we can evaluate statistically the universe of trees that should be considered appropriate for an analysis. We are still left with deciding which population model—described by gene flows, population size changes, and so on—best fits the set of gene trees.

Many of the approaches to Major Challenge 3 could use HECC now or are moving inexorably in that direction. Irrespective of scale, models that link geosphere and biosphere could be highly parameterized and, if they are, will ultimately rely on HECC when fully implemented. As noted in Chapter 3, geoscientists are beginning to couple climate modeling with environmental modeling and satellite data on ecosystem distribution to reconstruct the environmental history of Earth's ecosystems and to predict changes due to global warming. Concurrently, environmental modeling of species distributions is also becoming more common, and evolutionary biologists are beginning to use information about phylogenetic relationships to reconstruct the historical environmental envelopes of common ancestors down the tree. One can imagine the possibility of linking these classes of models and extending the coupled system farther back in time to examine how geological and climatological changes at a global scale might have influenced the distributions of species and biotas in terrestrial and marine environments. The complexity and precision of these reconstructions will depend on massive computational power. Such calculations will continue to stimulate advances in data integration and theoretical analysis, and the difficulties of these challenges will tax even the next generation of HECC.

Reconstructing how taxonomic elements of communities and ecosystems are assembled over time involves integrating the phylogenetic and spatial histories of many groups of organisms simultaneously. Current analytical approaches to this problem, conventionally undertaken on desktop computers, are widely regarded as inadequate because the simplifying assumptions of the methods and the models of spatial change are lacking in realism. Biogeographers, phylogeneticists, and computer scientists are collaborating to develop algorithmic approaches that will be able to extract the complex spatial and temporal histories of multiple groups simultaneously. Because of the large parameter space of the solution set and the algorithmic complexity, HECC will play a major role in data analysis.

From the outset, metagenomic studies have been intensely computational because they involve the assembly and comparisons of millions of gene fragments for hundreds or thousands of different kinds of microbes, many of which are new. The scale of such studies is expanding rapidly (see, for example, Rusch, 2007), and analyzing the results to address evolutionary questions, which will necessarily involve computationally intensive phylogenetic approaches, means that studies in evolutionary biology will more than ever need to push the frontiers of HECC. Metagenomic studies also reveal the remarkable diversity of protein families and their functional subdomains that have evolved. Predicting the functions of these domains is a highly complex computational problem that brings physical and chemical modeling together with evolutionary biology. No single algorithm has yet been established that works in all cases, but this is a subfield of very active research, with tools being developed and ongoing experimental validation of the predictions (Friedberg, 2006). The prediction of a protein's three-dimensional structure from its sequence is an area that already makes extensive use of HECC, and structure information could now be applied to understand the evolutionary diversification of protein structures and of families of functionally related proteins.

Simulations capable of representing the process of development, both qualitatively and quantitatively, are necessary for addressing Major Challenge 4. As noted in that section, developmental biologists are just beginning to create the algorithms, but progress is rapid, and they will soon be able to take advantage of HECC. Many of these models are based on relatively simple reaction-rate equations, but the number of parameters is very large. In many cases, parameters such as the rate of protein production and degradation, the rate of ligand diffusion, and the rate of receptor turnover can be specified only within certain limits. This means that the number of possible outcomes is too large to compute; instead, the space of all possible outcomes is sampled to gain an understanding of the developmental pathway.[2] Access to HECC will allow these developmental analyses to be done more efficiently and for a wider range of parameters. Likewise, the growing application of models from chemistry and physics, such as for the diffusion of ligands through the embryo (Gregor et al., 2007), has opened up the possibility of truly predictive models of development, which in turn promise the opportunity to understand the evolutionary outcomes of changes to the system.

Also, as noted in the discussion surrounding Major Challenge 4, analysis of gene regulatory networks and protein interaction networks is an important tool for understanding the development and evolution of phenotypes, and the analysis of both sorts of network can exhibit complexity such as will require HECC. For instance, it will be important to test how perturbations affect all manner of developmental events, and this requires multiple large-scale simulations. Also, as we develop a more detailed understanding of these networks and their effects on phenotypes, research will need to include more pathways and more organisms, thereby necessitating increased computational capability.

As noted in the discussion of Major Challenge 5, computational approaches are beginning to be used to recouple ecological and evolutionary dynamics. To date, most research has focused on simple cases, such as examining the dynamics within a single species with a very simplistic genetic architecture and genetic basis for phenotypic traits. Even so, such cases have contributed to some solid conceptual advances. Future success in this area, however, will depend heavily on analyses of models of communities of organisms with realistic ecological, genetic, and spatial structures, and the complexity of such models quickly brings us to the HECC domain.

The computational demands will be staggering, insofar as such analyses will involve the simultaneous tracking of thousands of genes in hundreds of interacting species, each with its own independent evolutionary history and genetic basis for traits. Extending this work to incorporate spatially explicit landscapes will further escalate the computational demands. One can imagine adding yet another set of equations to Lewontin's (Major Challenge 5) that would capture the complex trajectories of development (linking genotype to phenotype). Incorporating this last feature would truly integrate ecology and evolution by demonstrating the development of phenotypes from genotypes. At least for some model organism systems (sea urchins, *Drosophila*), developmental biologists have formulated equations that allow predicting phenotypic properties from the structure of complex gene networks (see Major Challenge 4). Very little has been done to tie such developmental predictions quantitatively to evolution or ecology (Kingsolver et al., 2007), no doubt because of the computational challenges, but this is clearly the direction in which coupled models of ecological and evolutionary dynamics are heading.

The computational methods that are essential to addressing Major Challenge 6 are largely manageable today, as explained in that section. As described there, evolutionary comparisons play a critical role

---

[2]A good example of such an approach is seen for models of the *Drosophila* segment polarity network, which maintains a segmental pattern in the early embryo. Computational analysis led to the surprising result that the network was remarkably robust in the face of environmental and genetic perturbations (van Dassow et al., 2000).

in genome annotation. In this way, we identify conserved protein coding regions and noncoding regions that presumably play a variety of regulatory roles. But the task of comparing and aligning genomes becomes increasingly difficult as we get more and more genomes and as we ask increasingly sophisticated questions. As this report is being written, not many bioinformaticians have routine access to very powerful computers, so it can be said that their research capability faces limitations. Algorithms and data are changing rapidly, and bioinformaticians often want to run their analyses repeatedly, tweaking parameters on each iteration. At this stage of development, then, it is advantageous for a researcher to work closely with his or her own machine, even if that constrains the scale of the calculations. As algorithms are perfected and databases swell, the need for HECC will grow rapidly. Making access fast and easy will also be critical in getting the user community to switch to state-of-the-art capabilities. These capabilities will also play an important role in developing gene models that can find and annotate genes independent of evolutionary conservation, which is essential when looking for genes that are evolving rapidly or are unique to specific lineages.

As our understanding of comparative genomics in wild populations improves, we will be better able to look for evolutionary signatures of selection and thus tease apart genome-level events that lead to speciation and macroevolutionary changes. While we have sophisticated theories and algorithms for this analysis, we are challenged to test these theories rigorously through large-scale genome analysis. As described above, transposable elements and the genome alterations they cause have played an important role in evolution, but piecing together the course of events is difficult from the computational standpoint. Ironically, the repeat structures created by transposable elements make the process of assembling whole genome sequences from raw data an even more complex computational problem. Right now, when we say a genome has been fully sequenced, that often applies only to its euchromatic region; the heterochromatic region, which often is rich with transposable elements, remains unassembled because computational methods are still lacking to make sense of the data.

While specific computational challenges and approaches in evolutionary biology have been discussed above, several additional observations can be made that apply across all levels of evolutionary research.

All of the biological sciences are data rich. This is not just in terms of volume per se but also in complexity, uniqueness or individuality, and their nonreducible characteristics. The data include such disparate materials as relatively simple DNA sequences, catalogs of museum specimens, photo images of collection materials, and movies of developing embryos. Thus the data storage, organization, and dissemination of biological material present challenges. (While the challenges of storing and making available genomic data are significant, it is even more difficult to store and share digitized visual data such as photographs, movies, and images of museum collections.) This flood of data has led to a need for computational tools and computer hardware for database storage, management, and usage. The torrent of biological data also has changed the evolutionary biology community's ability to study speciation. For example, earlier work on sympatric speciation used simple mathematical models with few parameters. These models are inadequate for addressing the complexity found with the new genetic data as well as more detailed information about population structure and dynamics. HECC will soon allow evolutionary biologists to move to large-scale simulations of individual-based, dispersed populations.

Through new technologies, especially for macromolecular analysis, biological investigations are becoming even more characterized by this data richness. In an era of high-throughput, high-information-content discovery in biological science, more and more research domains in evolutionary biology will be able to profit from high-end computing. To progress in exploiting the massive amounts of data, subdisciplines of evolutionary biology should strive to reach the point where HECC use is routine. Ultimately, this trajectory will unleash the potential for a very rich theoretical framework for evolutionary biology, just as exists today for the physical sciences. As this framework is built up and becomes robust, high-end computing will become pervasive within the community. In general, access to computing—using computing in the broadest sense—and at capabilities up to state of the art, combined with the data revolution, has already transformed studies in evolutionary biology, and it will grow more enabling as theory and experimentation continue their rapid advance.

Genomics and metagenomics data sets, individual genomes, and entire population or community genomes (metagenomics) all require the methods of computational evolution to gain understanding. For example, comparative analysis remains an essential tool for understanding biology. To utilize the explosion in genomics and metagenomics, efficient alignment methods and advanced statistical methods for characterizing sequence evolution are needed. Typically, a mathematical model (itself part of the framework of a specific model of evolution) is created to discern likely alignments. As ever-larger data sets are made available to the community, it may be possible to include correlation so that it no longer is necessary to treat each sequence position independently. Signal detection algorithms will need to be integrated into phylogenetic methods.

HECC is needed to cope with the data flow, which includes growing numbers of sequences, characters, numbers of species, and so on. Each parameter can take on thousands to hundreds of thousands of values, yet even more character data have to be added to gain confidence for resolving trees, whether those trees depict patterns of species interrelationships or the evolutionary patterns of genes within populations.

The data richness of biology often leads biologists to simplify their questions to the level at which they can be addressed in a reasonable amount of time on local computing resources—for example, reducing the number of parameters in modeling an ecological network. Enabling evolutionary biologists to readily exploit supercomputing power would significantly change the aims and scope of many research programs. Even today we can see how progress is limited by the relative scarcity of substantial computing resources at the high end: Workstations require months to solve medium-sized problems for modeling molecular evolutionary change even though new algorithms have provided some improvements. For theory, experimentation, and modeling to work coherently to advance evolutionary biology, computing must be available to sustain correlated intellectual inquiry. Long compute times are especially limiting when the range of models required is so large and the field has to depend on models and their validation, since no exact analytical solutions are possible.

The trajectory of the phylogenetics community as it seeks to probe the history of life is necessarily aimed at building and understanding ever larger trees. Thus another opportunity exists if we can encourage the development of community codes and community efforts to advance our understanding of the tree of life. An immediate example is the NSF's Assembling the Tree of Life project, which has nucleated a closely connected community. That community has the incentive and the common purpose to work together effectively to develop codes and use HECC extensively to improve models and their validation for the tree of life. Many other biological computing applications today are developed locally. Colleagues ask to use them, and at a certain point there is enough interest such that it is worthwhile investing resources into hardening them, bringing them up to standards that software engineers can live with, and making them more user-friendly. This bottom-up method of developing software works best for biology, where there is such a diversity of questions being asked, but work on developing, distributing and—especially—making it possible for these programs to work together is not well supported by the current funding mechanisms.

In addition, greater access to advanced computing environments can forge extensive partnerships with mathematicians and computer scientists, leading to clearer definition of computational problems and establishment of new algorithms. Greater collaboration with the mathematical, computer science, and engineering communities, as has happened in climate change and environmental biology, would greatly accelerate progress in evolutionary biology. These collaborations can also lead to capabilities for advanced visualization, methods for the implementation and application of evolutionary models, and capabilities for the interactive analysis of large, very complex data sets, all tailored to the particular needs of evolutionary biologists. These examples illustrate the many practical advantages of such access; the incentives for using advanced computing encompass more than just the classic NP-complete nature of generating and validating phylogenetic trees. But many steps must be taken before this vision can be realized.

The benefits of such access go beyond the ability to develop and use heuristic and approximate solutions for ever larger phylogenetic trees to advance our understanding of the history of life. They include how to apply this knowledge for the good of society. A deep understanding of evolution integrated into the fabric of biology provides the basis for all our understanding and knowledge of life and how living systems function. That, in turn, allows biology to contribute to developing new vaccines, antibiotics, and other medicines, predicting drug targets, managing natural resources, providing biosecurity through identification of pathogens and invasive species, and so on.

Evolutionary biology also presents challenges for the scientific computing community. That community has its roots more in the computational problems arising in physics, such as fluid flow, structural analyses, and molecular dynamics, and it has built up expertise and software for those sorts of problems. But many biological applications involve irregular data structures and unpredictable memory accesses (because the data come from strings, lists, trees, and networks), which place more demands on integer performance. So the community of expert algorithmicists and code builders must also be built up if evolutionary biology is to replicate the computational successes of the physical sciences.

What would evolutionary biologists need in a HECC facility? At the very least, the community needs uniform access to large data sets—tera- to petabyte scale for image data, giga- to terabyte scale for genomic data—and a network infrastructure that allows remote access and sharing. More generally, the challenge of large, interrelated data sets is a new one for biology, and the research community does not yet have the habit of looking for patterns in those data or the theoretical framework for doing so. When it does, we will be able to ask entirely new questions.

REFERENCES

Averof, M., and N.H. Patel. 1997. Crustacean appendage evolution associated with changes in *Hox* gene expression. *Nature* 388: 682-686.

Bader, D.A. 2004. Computational biology and high-performance computing. *Communications of the ACM* 47(11): 34-41.

Bader, D.A., B.M.E. Moret, and L. Vawter. 2001. Industrial applications of high-performance computing for phylogeny reconstruction. In Siegel, Howard J. (ed.), *Commercial Applications for High-Performance Computing*. Bellingham, Wash.: SPIE, 159-168.

Bader, David A., Allan Snavely, and Gwen Jacobs. 2006. *Petascale Computing in the Biological Sciences*. National Science Foundation Workshop Report. Arlington, Va., August 29-30.

Barton, N.H., D.E.G. Briggs, J.A. Eisen, D.B. Goldstein, and N.H. Patel. 2007. *Evolution*. Cold Spring Harbor Laboratory Press.

Batzer, M., and P.L. Deininger. 2001. Alu repeats and human genomic diversity. *Nature Reviews in Genetics* 3: 370-379.

Beaumont, M.A., and B. Rannala. 2004. The Bayesian revolution in genetics. *Nature Reviews in Genetics* 5: 251-261.

Beaumont, M.A., W. Zhang, and D.J. Balding. 2002. Approximate Bayesian computation in population genetics. *Genetics* 162: 2025-2035.

Brosius, J. 1999. Genomes were forged by massive bombardments with retroelements and retrosequences. *Genetica* 107: 209-238.

Carroll, S.B. 2005. *Endless Forms Most Beautiful*. New York, N.Y.: W.W. Norton.

Chen, L., A.L. DeVries, and C.-H.C. Cheng. 1997. Evolution of antifreeze glycoprotein gene from a trypsinogen gene in Antarctic notothenioid fish. *Proceedings of the National Academy of Sciences* 94: 3811-3816.

Coyne, J.A., and H.A. Orr. 2004. *Speciation.* Sunderland, Mass.: Sinauer Associates.

Cracraft, J., and M.J. Donoghue (eds.). 2004. *Assembling the Tree of Life.* New York, N.Y.: Oxford University Press.

Cracraft, J., M.J. Donoghue, J. Dragoo, D. Hillis, and T. Yates (eds.). 2002. Assembling the Tree of Life: Harnessing life's history to benefit science and society. Brochure produced for the National Science Foundation. Available at http://www.nsf.gov/bio/pubs/reports/atol.pdf/.

Deininger, P.L., and M.A. Batzer. 2002. Mammalian retroelements. *Genome Research* 12: 1455-1465.

Dobzhansky, T. 1964. Biology, molecular and organismic. *American Zoologist* 4(November): 49.

Felsenstein, J. 1988. Phylogenies from molecular sequences: Inference and reliability. *Annual Review of Genetics* 22: 521-565.

Felsenstein, J. 2004. *Inferring Phylogenies.* Sunderland, Mass.: Sinauer Associates.

Friedberg, I. 2006. Automated function prediction: The genomic challenge. *Briefings in Bioinformatics* 7(3): 225-242.

Gavrilets, S. 2003. Perspective: Models of speciation: What have we learned in 40 years? *Evolution* 57: 2197-2215.

Gregor T., E.F. Wieschaus, A.P. McGregor, W. Bialek, and W. Tank. 2007. Stability and nuclear dynamics of the bicoid morphogen gradient. *Cell* 130 141-152.

Hey, J., and R. Nielsen. 2007. Integration within the Felsenstein equation for improved Markov chain Monte Carlo methods in population genetics. *Proceedings of the National Academy of Sciences* 104: 2785-2790.

Huelsenbeck, J.P., F. Ronquist, R. Nielsen, and J.P. Bollback. 2001. Bayesian inference of phylogeny and its impact on evolutionary biology. *Science* 294: 2310-2314.

Kingsolver, J.G., K.R. Massie, J. G. Shlichta, M.H. Smith, G.J. Ragland, and R. Gomulkiewicz. 2007. Relating environmental variation to selection on reaction norms: An experimental test. *American Naturalist* 169: 163-174.

Levine, M., and E.H. Davidson. 2005. Gene regulatory networks for development. *Proceedings of the National Academy of Sciences* 102: 4936-4942.

Lewontin, R.C. 1979. Fitness, survival, and optimality. Pages 3-21 in D.H. Horn, R. Mitchell, and G.R. Stairs, eds. *Analysis of Ecological Systems.* Columbus, Ohio: Ohio State University Press.

Lewontin, R.C. 2002. Directions in evolutionary biology. *Annual Review of Genetics* 36: 1-18.

Li, W.-H. 1997. *Molecular Evolution.* Sunderland, Mass.: Sinauer Associates.

Lindström, J., and H. Kokko. 2002. Cohort effects and population dynamics. *Ecology Letters* 5: 338-344.

Lynch, M., and J.S. Conery. 2003. The origins of genome complexity. *Science* 302: 1401-1404.

Lynch, M. 2007. *The Origins of Genome Architecture.* Sunderland, Mass.: Sinauer Associates.

Maynard-Smith, J. 1982. *Evolution and the Theory of Games.* Cambridge, England: Cambridge University Press.

Meagher, T.R., and D.J. Futuyma (eds.). 2001. Evolution, science, and society: Evolutionary biology and the national research agenda. *American Naturalist* 158 (Supplement): 1-46. Available at http://www.journals.uchicago.edu/ASN/meagher.html/, and at http://evonet.sdsc.edu/evoscisociety/.

Nielsen, R. (ed.). 2005. *Statistic methods in molecular evolution.* New York, N.Y.: Springer Verlag.

Nowak, Martin A. 2006. *Evolutionary Dynamics: Exploring the Equations of Life.* Cambridge, Mass.: Harvard University Press.

NRC (National Research Council). 1995. *Effects of Past Global Change on Life.* Washington, D.C.: National Academy Press.

NRC. 2007. *The New Science of Metagenomics: Revealing the Secrets of Our Microbial Planet.* Washington, D.C.: The National Academies Press.

NSF (National Science Foundation). 1998. *Frontiers in Population Biology.* Workshop report from the Population Biology Task Force. Available at http://www.nsf.gov/publications/ pub_summ.jsp?ods_key=biorpt1098.

NSF. 2005a. *Frontiers in Evolutionary Biology.* Workshop report, Document number biorpt080106. Available at http://www.nsf.gov/publications/ods/results.cfm?url_type=Reports&url_subtype=Biology&browse_type=org_type.

NSF. 2005b. *Assembling the Tree of Life.* Multiple workshop reports available at http://www.nsf.gov/publications/ods/results.cfm?url_type=Reports&url_subtype=Biology&browse_type=org_type.

Ricklefs, R.E., and D. Schluter (eds.). 1993. *Species Diversity in Ecological Communities.* Chicago, Ill.: University of Chicago Press.

Riesenfeld, C.S., P.D. Schlos, and J. Handelsman. 2004. Metagenomics: Genomic analysis of microbial communities. *Annual Review of Genetics* 38: 525-552.

Rusch, D.B., A.L. Halpern, G. Sutton, K.B. Heidelberg, S. Williamson, et al. 2007. The *Sorcerer II* Global Ocean Sampling expedition: Northwest Atlantic through eastern tropical Pacific. *PLoS Biol* 5(3): e77. doi:10.1371/journal.pbio.0050077.

Schlosser, G., and G.P. Wagner (eds.). 2004. *Modularity in Development and Evolution.* Chicago, Ill.: University of Chicago Press.

Shapiro, M.D, M.E. Marks, C.L. Peichel, B.K. Blackman, K.S.Nereng, B. Jónsson, D. Schluter, and D.M. Kingsley. 2004. Genetic and developmental basis of evolutionary pelvic reduction in three spine sticklebacks. *Nature* 428: 717-723.

Stern, D.L. 1998. A role of Ultrabithorax in morphological differences between Drosophila species. *Nature* 396: 463-466.

Tyson, G.W., J. Chapman, P. Hugenholtz, E.E. Allen, R.J. Ram, P.M. Richardson, V.V. Solovyev, E.M. Rudin, D.S. Rokhsar, and J.F. Banfield. 2004. Community structure and metabolism through reconstruction of microbial genomes from the environment. *Nature* 428: 37-43.

van Dassow, G., E. Meir, E.M. Munro, and G.M. Odell. 2000. The segmenta polarity network is a robust developmental module. *Nature* 406: 188-192.

Vincent, T.L., and J.S. Brown. 2005. *Evolutionary Game Theory, Natural Selection, and Darwinian Dynamics*. Cambridge, England: Cambridge University Press.

Whitlock, M.C., and R. Gomulkiewicz. 2005. Probability of fixation in a heterogeneous environment. *Genetics* 171: 1407-1417.

Wilmers, C.C., E. Post, and A. Hastings. 2007. A perfect storm: The combined effects on population fluctuations of autocorrelated environmental noise, age structure, and density dependence. *American Naturalist* 169: 673-683.

Yang, Z. 2006. *Computational Molecular Evolution*. Oxford, England: Oxford University Press.

5

The Potential Impact of HECC in Chemical Separations

INTRODUCTION

Separation processes—the production of two or more streams that are different in composition from that of the feedstock—are ubiquitous. They can operate on the smallest amounts of matter that consist of more than one atomic or molecular species or on the scale of the cosmos, where atoms and subatomic fragments are separated by the action of gravitational and other fields. Around the world, separation processes are building blocks for a wide range of industrial and environmental processes that impact society broadly and in many ways. For example, chemical separations are essential for the following purposes:

- Removal of toxic substances like mercury from the flue gases of coal-fired power plants and removal of a range of organic and inorganic pollutants from wastewater streams.
- Removal of the greenhouse gas carbon dioxide from power plant flue gases.
- Recovery of very dilute but highly radioactive cesium-137 from nuclear-waste streams (NRC, 2000).
- Separation of nitrogen, carbon dioxide, water, and other contaminants in gas from natural gas wells, coal bed methane wells, and landfills so that the methane can be added to the interstate pipeline system.
- New separation applications to accommodate the commercialization of green products.
- Production of potable water in many developing countries.
- Purification of a growing number of new drugs from their chiral (mirror-image) compounds, which can in many instances be highly toxic.

The energy requirements to achieve the separated products are substantial, however, and come at a time when we can least afford it, with sometimes negative environmental consequences that can no longer be ignored. In the mid-1990s, separation processes in the chemical industry alone consumed about 7 percent of the total energy used in the United States (NRC, 1998); separation processes used in other industries, while difficult to quantify from an energy-use standpoint, probably added one to several percentage points to that number. About 60 percent of the total energy requirements of the chemical and petroleum processing industries are consumed by separation processes (DOE, 2005). Capital investments in separation processes are also a very important factor, with 40-70 percent of the total investments in various separation-intensive industries being consumed by these processes (Humphrey and Keller, 1997). Given that separation processes consume so much energy, it is clear that they also contribute very significantly to the nation's output of greenhouse gases. Thus for three reasons—energy use, investment costs, and environmental considerations—the incentives to improve these processes, as well as to invent and develop new ones, are very great.

Box 5-1 portrays both the breadth of the separations field and the large number of disparate industries in which these processes are applied. Most chemical separation processes are based on thermodynamic equilibrium considerations. When, for example, a liquid stream containing two or more components is heated and forms a vapor phase in contact with the liquid, at least a partial separation of the components is possible if the resulting two phases at equilibrium have different compositions. Distillation is highly effective at separating compounds based on differences in their relative volatilities. From a design point of view, distillation-based processes are favored not only because their mechanical simplicity often leads to low investment costs but also because their design requires a much smaller set of phase-equilibrium data than all other separation options to quantify and optimize the efficiency of the separation. This fact accounts in large part for the historic preference for distillation over alternative methods.

Distillation, because it requires that the mixture be repeatedly vaporized and condensed, nonetheless consumes tremendous amounts of energy. Historically, energy consumption and its concomitant carbon dioxide release were not deemed to be of great concern, so chemical industries tended to design

---

BOX 5-1
Major Separation Processes and Industries
That Depend Heavily on Chemical Separations

Separation Processes

Distillation	Membrane-based	Filtration
Solvent extraction	crystallization	Bubble/foam fractionation
Supercritical gas extraction	Ion exchange	Electrodialysis
Gas and liquid adsorptions	Drying	Liquid chromatography
Gas absorption		

Industries Served

Organic and inorganic chemical production	Electronic products	Industrial, municipal, and agricultural waste treatment
Polymer production	Food processing	
Petroleum refining	Biochemical products	
Pharmaceutical production	Biofuels production	Hospitals and other health-care entities
Ore, coal, oil, and gas extraction and cleanup	Advanced biotech products	Homeland security

TABLE 5-1 Examples of Mass-Separating Agents and Their Applications

Mass-Separating Agent	Application
Zeolite molecular sieve	Oxygen from air, hydrogen recovery, isomer separations, glucose-fructose separation, CO_2 removal from gas streams, water removal from ethanol
Activated carbon	Removal of trace organics from water and air, of color from petroleum fractions, and of odor and taste bodies from water
Ion exchange resin	Removal of specific ions from various, usually aqueous, streams
Functionalized solvent	Separation of derivatized organics from simpler organics
Water	Separation of ions and polar organics from organic phases
Polymer membrane	Nitrogen from air, hydrogen recovery, water removal from gases, water purification, CO_2 recovery, desalination, biological materials separations
Filter	Removal of solids from gases and liquids
Flocculating agent	Concentration of fine particles and biological agents in aqueous streams

separation systems based on distillation if it was a viable option, turning to other options only if it was not. This approach remains dominant, even though most of the alternatives to distillation would require less energy and produce less CO_2. Given that distillation is by far the most common separation process, used in as much as 80 percent of all the chemical separations listed in Box 5.1, optimization of phase equilibria will remain an important grand challenge for the chemical separations industry.

It is also true that distillation is sometimes not an effective option. Instead, mass-separating agents (MSAs)—solvents, absorbents, adsorbents, membranes, and so on—are often added to amplify the separating capability for these more intractable systems, while potentially providing for more economical, environment-friendly solutions. Some examples of MSA-based processes are given in Table 5-1, which we amplify by focusing on two examples of their use that have broad societal implications.

Example 1: Pure Oxygen from Air

Even though oxygen is already produced inexpensively on a massive scale, the number of uses and overall volume produced could grow substantially if its price were cut even more. Some of the existing and potential applications include the following:

- Feeding oxygen instead of air to power plant furnaces to reduce the volume of flue gas produced and to increase the percentage of carbon dioxide, sulfur oxides, and nitrogen oxides in the flue gas, dramatically reducing the cost of their recovery. Whether this use comes about is highly dependent on the need to sequester the carbon dioxide.
- Feeding oxygen to gasification reactions such as occur in next-generation, integrated gasification combined cycle (IGCC) coal-based power plants, which may be the wave of the future.
- Feeding oxygen instead of air to aerobic waste-treatment processes, thereby reducing equipment size and costs.
- Feeding oxygen to a large number of organic oxidation processes to improve selectivities and reduce energy costs.

The savings would have to be larger than the capital and energy costs of producing the oxygen in order for these applications to be realized and grow. The secret to lowering oxygen costs would appear to be the discovery of new MSAs that can operate at temperatures and pressures that make it possible to use waste heat as the energy source. This strategy is already being tested in an IGCC pilot plant with a new technology—a ceramic membrane operating at about 800°C—which makes it possible to use waste heat from the furnace as the sole energy source for the air separation. The development of other MSAs that could operate in a similar fashion for other processes could usher in a revolution in oxygen use.

Example 2: Separations of Chiral Compounds

Chiral compounds are isomers whose structures cannot be superimposed on each other. Because they are similar in chemical composition to their mirror-image relatives, their thermodynamic properties such as vapor pressure, solubilities, and other properties are quite similar, making it often very difficult to separate them. The most common separation technique is to seed a melt of the two isomers with a crystal of one of them, causing that isomer to precipitate out of solution preferentially. Furthermore, it is almost always the case that, in nonbiological syntheses at least, chiral isomers are produced in equal amounts. Unfortunately, biological systems such as the human body do react differently to the two isomers, sometimes dramatically. For example, thalidomide, $C_{13}H_{10}N_2O_4$, a chiral isomer and a sedative, produced major birth defects when the product being sold contained more than a negligible amount of its chiral twin, which produced those defects. Nor is this an isolated problem. Naproxen, a popular pain reliever today, has a chiral twin that is a liver toxin.

More and more, the fraction of new drugs coming on the market that are chiral is growing, and they must undergo precise and virtually complete separations from their chiral twins to eliminate the possibility that these twins might produce unfortunate side effects. What are needed are MSAs that can precisely separate chiral isomers and make it possible to produce drugs of the proper chiral purity much more easily and cheaply. However, even though separation systems that rely on the use of MSAs are an important area of growth in the separations industry, the design of new MSA systems is severely hindered by the lack of physical property data and novel design leads.

MAJOR CHALLENGES FACING CHEMICAL SEPARATIONS

There are three major challenges facing those concerned with the development of efficient chemical separations:

1. How can we predict physical properties accurately enough to set the optimal conditions for separating mixtures using distillation and MSA materials?
2. How can we design, construct, and mass produce MSAs with appropriately engineered three-dimensional structures (when appropriate) that make it easier to do difficult separations rapidly and efficiently?
3. How can we design overall separation systems that incorporate several individual separation units for economically optimal separations of complex mixtures?

This list is based on several documents developed in recent years by the chemical separations community. An NRC report (1998) was used as the starting point. Reports from the Chemical Industry Vision 2020 Technology Partnership[1] and another NRC report (2003), which examined the broader question of computational chemistry and materials science, provided insights into new challenges that are apparent

---

[1] See http://www.chemicalvision2020.org/library.html. Last accessed on July 25, 2008.

now but were not at the forefront at the time of the 1998 NRC study. Finally, the committee benefited from presentations by Joan Brennecke (University of Notre Dame), Anne Chaka (National Institute of Standards and Technology), and Jeffery Siirola (Eastman Chemical) at a small workshop in December 2006 (see Appendix B). From these sources, the committee developed the three major challenges listed above.

Dimensions, operating temperature, and operating pressure of the individual units are determined by vapor-liquid equilibrium data, the focus of Major Challenge 1, for both traditional distillation and for optimizing new MSA materials. Major Challenge 2 focuses on determining the appropriate nanoscale structures when MSAs are used and on the ability to predict interactions of the MSA components with the chemicals to be separated. Major Challenge 3 is an overall operations research problem that deals with optimizing the interplay of multiple separations processes to achieve high-performance separations systems of complex mixtures. It is important to emphasize that the three major challenges pertain at very different spatial scales. Major Challenges 1 and 2 require a better understanding of molecular interactions within gases, liquids, and solids; Major Challenge 2 also deals with connecting particular nanoscale characteristics to the manufacturing of engineered separation materials with dimensions on the order of micrometers to meters. Finally, Major Challenge 3 addresses the design of systems on the scale of tens of meters.

As progress is made on Major Challenges 1 to 3, we will be able to enlarge the space of options available for purification systems and make design decisions that are closer to optimal. Looking to the future, we will also create options for addressing critical separation technologies such as the following, for which no acceptable separation schemes currently exist:

- Efficient recovery of highly dilute species from solutions.
- Recovery of CO_2 from stack gas and automobile exhaust and sequestration of this compound.
- Development of less energy-intensive routes for producing oxygen.
- Efficient removal of sodium and other inorganics from water.
- Efficient separation of optical isomers to produce chirally pure products.

The major challenges of chemical separations are driven by the demand for the capabilities they can enable. This is in contrast to the situation in astrophysics, evolutionary biology, and some aspects of atmospheric science, where the motivation for overcoming the major challenges is the gap in understanding that must be filled to make scientific progress. The situation for chemical separations is much like that in operational meteorology because in both cases a capability already exists. Pushing the frontier amounts to improving and extending that capability, which in the case of chemical separations has staggering implications for our ability to assume prudent stewardship of our environment while maintaining our economic competitiveness. We now examine the three major challenges in chemical separations in more detail.

Major Challenge 1: Accurately Predicting Physical Properties for Phase Equilibria

How can we predict physical properties accurately enough to set the optimal conditions for separating mixtures using distillation and MSA materials? Most current separations are equilibrium based. For example, when a feedstock undergoes a phase change, the two resulting phases typically possess different compositions, and a chemical separation has been achieved. In other cases, an MSA can be equilibrated with a single phase and the MSA phase can selectively remove certain chemicals out of the original phase. If the MSA is selective, what remains of the original phase will have a new chemical composition.

Once again, a separation can be made. For systems such as these, the phase-equilibrium information is absolutely essential for determining the ease or difficulty of obtaining the desired separation and for predicting operating efficiencies and choosing among competing process designs. The field can utilize both data measured experimentally and data predicted from computational and theoretical chemistry methods that determine the energetics driving phase separations. Chemical separations is a field that benefits from combining experimental and computational approaches to the acquisition of physical property data, and this combined approach is anticipated to dominate for the foreseeable future.

For easy separations of binary mixtures such as water and ethylene glycol, no great accuracy is required. But for mixtures that are not thermodynamically ideal, such as acetic acid and water or isopropanol and water, for mixtures that have boiling point differences of 10°C or less, and for mixtures with three or more components, satisfactory definition of the phase equilibria can become an experimental nightmare, opening a primary role for computational predictions. Seldom do feed mixtures have just two components, and determining the appropriate mathematical representation of the experimental data escalates dramatically in difficulty as the number of components increases, which suggests that more research is needed for mathematical models and optimization. These predictions must be quite precise to determine the system specifications and the energy requirements for the separation.

So why is industry not aggressively attacking Major Challenge 1? One reason is that quite a few of the separation processes used in the petrochemical industry (which itself is a large fraction of the chemical industry) have historically been purchased from large engineering companies rather than developed in-house. Rather than pioneering a new process, the companies that operate these plants tend to think in terms of improving marketing, logistics, and supply chains as ways to differentiate themselves and increase their profitability. In addition, training and education in computational chemistry and mathematical optimization are not well-integrated into the chemical engineering and chemistry curriculum, thus limiting the extent to which methods and algorithms can inform optimality of phase equilibria and process operation.

However, it is likely that these accepted business practices will change in the near future if a green chemical revolution really takes hold. The separation requirements for a green product are normally greater than for a competing conventional product because of the greater complexity of green raw materials, the greater incidence of nonideal mixtures, and the desire to reduce the energy costs associated with those nonideal mixtures. The advantage of being first in the green market will probably drive some companies to begin addressing Major Challenge 1 as a way of entering that market while controlling costs. And once some companies have made that investment, the competitive landscape could shift quickly to favor the companies with stronger computational capabilities and resources.

Major Challenge 2: Designing and Producing MSAs for Difficult Separations

How can we design, construct, and mass produce MSAs with appropriately engineered three-dimensional structures (when appropriate) that make it easier to do difficult separations rapidly and efficiently? The chemical structure of an MSA will determine its physical properties, the nature and degree of interactions with other compounds, and, ultimately, its suitability for a given application. Accurate prediction of the properties of liquid MSAs used in extraction systems will allow engineers to design such systems. For the solid MSAs used in adsorption and membrane systems, the three-dimensional structure of the material is also of importance in determining its potential to accomplish the desired separation efficiently. We must also evaluate how amenable those potential MSA structures are to efficient and accurate production. The structures can be important not only because of their inherent thermodynamic equilibrium selectivity but also because of their various hydrodynamic and mass-transfer qualities, which can increase separation efficiencies. At present we have only a limited understanding of the relationship between structure and separation selectivity and efficiency.

As an example of unexploited opportunities from new MSAs, consider metalorganic framework materials (MOFs), a class of materials that is currently attracting much attention for their selective adsorption of a vast array of solutes arising from the large number of possible MOF structures. MOFs have potential for use in many different applications, including gas storage, separations, catalysis, and sensors.[2] MOFs have been shown to have some of the highest surface areas (for example, 4,900 m^2/g) reported for any material to date, making their separation capacities quite large. Experimental characterization of the most promising MOF materials—to understand how adsorption and diffusion of the materials to be separated correlate with structural features such as pore size, surface area, and void volume—is time consuming, and computational chemistry can suggest which novel structures are the most promising candidates for that experimental characterization. Computational chemistry in this case guides more rational design improvements for existing MSAs and de novo design of new MSAs for separations with less stringent accuracy requirements than those encompassed by Major Challenge 1.

Major Challenge 3: Designing Optimal Separation Systems with Multiple Separation Units

How can we design overall separation systems that incorporate several individual separation units for economically optimal separations of complex mixtures? Once a separation scheme has been proposed, determining the efficiency of the separation, the optimum operating conditions for each unit, and the sizing of the units shown and the connecting piping is rather straightforward if we have a good understanding of the physical properties of the chemicals or mixtures and of the structures and performance of any MSAs used in the process. Yet, for any desired separation, a tremendously large number of possible separation schemes exist, which might employ any combination of the processes described in Box 5-1. Determination of which processes to employ and in what order is the responsibility of the process engineer. The current state of the art relies heavily on the experience of the process engineer and on rough rules of thumb (for example, if distillation works, use it). The number of solutions to this problem is very large, with the best solution being influenced by value judgments on cost, waste produced, time required, safety, and so on.

Major Challenge 3 consists of two related but different problems. In the determination of the optimum process system to be used to make a given product (or system of products), one must (1) determine all possible systems to be considered and then (2) evaluate which of the available solutions is optimal based on the design criteria.

While both problems are difficult, the community's progress in systematically attacking the second problem seems more advanced. Many chemical engineering groups in industry and academia are strong in computational modeling of processes to enable their design, optimization, and control. Once a process engineer has roughed out a proposed multiscale system, the tools and expertise exist to analyze its performance and then optimize the design.

Unfortunately, before we can even talk about best solution for a desired chemical separation, we need a method for addressing the first problem, that of generating all the possible options to be evaluated. Currently, there is no clear algorithm for methodically surveying the space of design options, although the best process designers seem to have good intuition in this regard. For instance, an experienced process engineer can create multiple distinct processes for achieving some common separations, not all of which

---

[2]See, for example, Cho et al., 2006; Dinca et al., 2006; James, 2003; Kitagawa et al., 2004; Latroche et al., 2006; Matsuda et al., 2005; Millward and Yaghi, 2005; Mueller et al., 2006; Pan et al., 2006; Panella et al., 2006; Rosseinsky, 2004; Snurr et al., 2004; and Wang et al., 2002.

would be obvious to a novice (or an algorithm). If an algorithm could be developed to methodically scan the design space—presumably drawing on a combination of expert knowledge, machine learning, experimental data, and computational simulations—an optimal design could in principle then be found through local or global optimization techniques if sound models for cost functions were defined. The number of dimensions in the design space and the complexity of some of the component simulations suggest that many such system optimization problems could require HECC.

This is a long-term challenge for the industry and for computation. Little published research has appeared in this area, and the challenge remains at a very early stage of conceptualization. And there is, at present, little economic incentive to attack Major Challenge 3, because chemical companies are able to stay competitive without investing in new methods of process design. If appropriate algorithms can be developed that systematically create a library of options from which an optimum solution will be drawn, significant advances in the design and construction of efficient systems will surely be realized.

POTENTIAL IMPACTS OF HECC FOR CHEMICAL SEPARATIONS

There are numerous examples of computational chemistry leading to new understanding about the behavior of chemicals in separations systems. Examples are shown in Box 5-2. Computation can play the vital role of informing the experimental plan and focusing the expensive and time-consuming experiments on the precise set needed for a given design. Further, it has many potential advantages over the experimental evaluation of material properties, including these:

- Safe determination of properties for chemical species that are highly toxic or highly reactive, or both.
- Determination of properties for chemical species that have not yet been synthesized or purified.
- Rapid prediction of properties for a wide range of chemical compounds.
- Inexpensive determination of relative properties.
- Fast screening of potential solvents or MSAs.

Computational approaches have great potential for facilitating more progress on Major Challenges 1 and 2. Because simulations of molecules of industrial importance and of realistic systems are computationally demanding, it is likely that HECC resources will be required. HECC enables more accurate predictions of properties, which can lead to gains in efficiency and cost, whether through more precise design of thermal-energy-based separations (for example, distillation) or the use of an appropriate MSA. It also can expand the range of chemical and parameter options being evaluated. For example, a profoundly improved MSA for the selective removal of oxygen from air is much more likely to be discovered using a computational approach rather than an experimental one. Typically, design of thermal-energy-based separations or of new MSAs is based on the characteristics of the best-known materials available to date. Modest changes to the known chemistry might bring small improvements in performance, but such incremental approaches will not bring true breakthrough technology. Breakthroughs are likely to come when examining some entirely unexplored region of parameter space. HECC is ideally suited for mapping wide expanses of parameter space and highlighting potentially exciting regions. The HECC-developed map can then be used to direct the design of next-generation MSAs ripe for experimental evaluation.

Major Challenge 3 would be critically dependent on HECC if the cost model and parameter space can be defined so that well-developed optimization and machine learning algorithms can be applied. But, as noted above, there is little movement to attack that challenge.

BOX 5.2
Examples of Computational Chemistry Enabling Improved Understanding of the Behavior of Chemicals in Separations Systems

- *Phase behavior.* Prediction of phase behavior of water and small organic molecules. This understanding is essential for the design of separation systems based on thermodynamic interactions, such as distillation and extraction. See, for example, Rablen et al., 1998.
- *Drug design.* The design of new pharmaceuticals has been significantly impacted by advancements in computational chemistry. Today, computational chemistry can help predict the specific structures of new pharmaceuticals based on the molecular properties of specific target molecules (Jorgensen, 2004). Additionally, it can be used to refine structures which have been discovered using experimental methods. This coupling of computation and experimentation can lead to new materials with superior properties (Martin et al., 1993).
- *Materials screening.* Computational chemistry provides a method for fast and effective screening methods for new drugs, catalysts, and materials. It allows for evaluation of a much broader phase-space than would be possible with experiments alone (Walters et al., 1998).
- *Materials design.* Design of new solid structures capable of use as MSAs. See, for example, Lipkowitz (1998).
- *Design of microporous solids.* The design of microporous solids with controlled pore size, volume, and surface area is of tremendous importance in fields such as adsorption and catalysis. Férey et al. (2005) describe the use of targeted chemistry and computational design to create a crystal structure with very large pore size and surface area.
- *Industrial success stories.* Westmoreland et al. (2002) cite the following as examples of notable industrial successes in the use of computational chemistry:
  — Rhône-Poulenc used quantum mechanical calculations of a Flory χ-parameter and relative reactivities in developing an antiscratch additive for polyurethane coatings.
  — Rhône-Poulenc used computation to determine that it would not be possible to develop a material to compete with its competitor using a nylon basis, a valuable negative result.
  — Lubrizol used a QSPR model for gasoline additive formulation to reduce testing costs by one-third for predicting intake valve deposits in BMW, Ford, and Honda engines.
  — Dow estimated that each ΔH_f calculation saved the company $50,000 in testing costs in 1996 and over $100,000 in 2000.
  — Mitsubishi Chemicals reports that 5 percent of the patents from its Yokohama facility involve some computational modeling.

To successfully address Major Challenges 1 and 2 requires building on the capabilities of computational chemistry, which now include calculations at the molecular scale with algorithms based on quantum chemical theories and classical analogs that evaluate the energetics of molecular conformations, as well as statistical mechanical methods that sample those conformations consistent with thermodynamic variables such as temperature and pressure. The general strategy typically employed in computational chemistry is to combine these methods based on the following diagram:

Quantum chemistry calculations → Classical empirical force fields
↘   ↙
Statistical mechanical sampling
↓
Phase diagram

The primary uses of quantum chemistry calculations are to calculate for model chemical compounds the relative energies of different molecular conformations—the charge density descriptions from which one derives partial atomic charges—and the intermolecular interaction energies. All of these calculated quantities can then be used to develop the empirical force fields, which can approximate the classical forces on all the atoms or molecules being studied. Because this force-field calculation is less expensive computationally than a full-fledged quantum calculation of the entire system, it is the more feasible approach for larger and more complicated systems. Furthermore, it is assumed that the quantum mechanical results for a given compound are still valid when that compound is part of a larger molecule, although it is recognized that this is not always a good approximation. This problem of transferability is most limiting when electronic structure calculations are transformed into empirical classical force fields. In either case, once the quantum or classical results have been validated against experimental data, the resulting energetic models of an MSA material or of the chemicals in a thermal-energy-based unit process feed into statistical mechanical simulations. Statistical mechanics provides a solid theoretical foundation for defining equilibrium and dynamical sampling schemes of these molecular conformations, thus allowing the generation of a global minimum structure, a phase diagram, absorption probabilities, or transport properties such as diffusion, all of which are needed by the engineer or scientist intent on developing new chemical separation schemes.

A sweet spot for such methods at present is when qualitative predictions suffice for identifying phase equilibria thermodynamic parameters or promising MSAs to investigate experimentally. For example, using molecular simulations, the nitrogen adsorption preferences within selected MOF materials known as IRMOF-1 and IRMOF-16, shown in Figure 5-1, were predicted. The calculations predicted that nitrogen prefers to associate with only the corner regions of IRMOF-1, while for IRMOF-16 it associates with not only the corners but also the faces of the benzene rings. Thus, experimental efforts would be steered toward IRMOF-16 because it is predicted to have greater nitrogen adsorption rates and capacities. When successfully executed, such computational modeling can direct experimental programs so that highly effective MSAs can be produced with a minimum of time-consuming experimentation.

In other cases, qualitative insight is not enough and quantitative predictions are necessary. In order for computational chemistry to develop predictive capabilities good enough to overcome Major Challenges 1 and 2, the following are needed:

- Scalable algorithms for quantum electronic structure calculations.
- Greatly improved classical force-field accuracy.
- Improved statistical sampling via molecular dynamics and Monte Carlo methods.
- Extensive validation studies on resulting phase equilibria and MSA structures.
- Training and education of the next generation of computational chemists and chemical engineers.

FIGURE 5-1 Grand Canonical Monte Carlo simulation at 77 K of the adsorption of nitrogen within the unit cell of two MOF materials. IRMOF-1 (top) is represented by metal atoms at the corners of the unit cell connected by a [linker structure]-linker. IRMOF-16 (bottom) consists of metal atoms at the corners of the unit cell connected by a [linker structure]-linker. The calculations show that the different MOF materials have different capacities for nitrogen separation depending on the linker chemistries. While nitrogen is concentrated only at the metal sites of IRMOF-1, it can absorb at both the metal sites and linker aromatic rings in IRMOF-16. These calculations suggest further MSA designs without costly experimentation. Red is higher density—that is, the molecule was found at that location a relatively large number of times over the course of the simulation—with orange denoting a lower density, yellow being lower still, and blue signifying that no molecules were predicted at those locations.

CURRENT FRONTIERS OF HECC FOR CHEMICAL SEPARATIONS

Algorithms for Quantum Electronic Structure Calculations

The 1998 Nobel prize in chemistry went to John Pople for his development of computational methods in quantum chemistry, including the mean field approximation of Hartree-Fock (HF) methods and electron correlation methods that enable increasing levels of accuracy, and to Walter Kohn for his development of an alternative approach to electronic structure, known as density-functional theory (DFT). The Nobel prize press release emphasized that "as well as producing quantitative information on molecules and their interactions, [computational chemistry] also affords deeper understanding of molecular processes that cannot be obtained from experiments alone."[3]

Quantum chemistry is based on reformulating the Schrödinger eigenvalue equation into a large set of algebraic equations expanded in some convenient mathematical basis set—typically Gaussianfunctions—and the development of well-defined approximations to electron-electron interaction potentials. HF uses a mean-field approximation to treat electron interactions, such that the computational complexity scales quadratically with molecular size, although the algebraic steps of matrix diagonalization scales cubically to dominate the computational complexity for a large system. DFT offers the advantage of similar computational scaling complexity while treating electron correlation beyond the HF mean-field approximation. It is important to emphasize that because DFT captures only certain types of electron correlation, the quality of DFT calculations is still under debate. This is especially the case for weak nonbonded interactions, and the development of new DFT functionals is an active area of research. Routine calculations for these methods are now completely feasible for molecules with hundreds of atoms, and heroic calculations for ~1,000 atoms are possible on the most powerful computers and with a good deal of computing time.

A feasible and often more robust alternative to post-HF methods is the Moller-Plesset Perturbation (MP2) series to describe electron correlation beyond the mean-field HF reference. MP2 refers to the mathematical model that perturbs the HF reference to include electron correlations up to second order. The MP2 method scales with the 5th power of system size because the formulation of MP2 uses delocalized molecular orbitals that arise from standard HF calculations. However, the molecular orbitals can be localized, and there has been a great deal of progress toward developing a "local-MP2" method that scales only quadratically with molecular size and comes to within a few percent of reproducing the exact MP2 energy for a given basis set, making the computation of molecules comprising hundreds of atoms completely feasible.

The gold standard of quantum chemistry calculations is coupled cluster methods, a general formulation with high levels of electron correlation that can use any orbital reference. While these theoretical models have been formulated into algorithms, they have severe scaling requirements (scaling at least with the 7th power of system size), which have traditionally limited their applications to very small system sizes (tens of atoms).

The post-HF methods provide a good to very good level of accuracy with regard to relative conformational stabilities and barriers, charge densities, and weak intermolecular interactions. They provide excellent input for developing empirical force fields for many classes of chemical compounds. HECC will make it less expensive to perform electronic structure calculations and will enable the calculation for much larger molecules of importance when the physics is well described by HF/DFT or MP2 levels of theory, on the condition that these algorithms can be deployed on massively parallel architectures, which is a limiting factor since the algorithms are still only weakly parallelizable. For classes of more complex materials, current capabilities of these methods may themselves inherently limit the accurate calculation of phase equilibria data, and coupled cluster methods are to be preferred.

Improved Accuracy of Molecular Mechanics Force Fields

Empirical force fields derived from electronic structure calculations and experimental data, coupled to classical molecular dynamics or Monte Carlo sampling schemes, are the main component of all computational studies of materials chemistry to date. Overall, they can be the weak link in accurate

---

[3] Available at http://nobelprize.org/nobel_prizes/chemistry/laureates/1998/press.html.

determination of phase equilibria data through computation if they have not been sufficiently validated against experiment. For example, the physical properties of homogeneous liquids of certain classes of molecules, such as Lennard-Jones fluids or liquid water, are qualitatively and even quantitatively well described by empirical molecular mechanics force fields based on productive collaborative work between theory and experiment over the last several decades. The challenge of using empirical force fields in the chemical separations industry is that each new thermal-based separation or MSA material requires empirical force field development based on an affordable electronic structure calculation, and experimental validation is then required in order for those force fields to be usefully deployed.

The challenge in constructing a force field from quantum calculations lies in determining the form of the mathematical function. One must strike a balance between enough complexity to accurately describe the fundamental interactions of matter on the one hand and simplifications that decrease the computational complexity on the other. For instance, approximate classical models represent bonds and angles as harmonic springs, dihedral angle conformations by a truncated Fourier series, pairwise non-bonded interactions with a Lennard-Jones function, and electrostatic interactions between point charges by Coulomb's law. There are several empirical force fields of this type in use, and they are widely used in industry and academic research settings.

Beyond the less-accurate two-body potentials described above, the most recent generation of empirical energy functions incorporates the many-body effects of polarizability by modeling how the electron density responds to an electric field that is generated by the condensed phase of a material of interest. It is generally agreed that including polarizability into empirical force fields is necessary for good quantitative agreement between simulations and experiments not at ambient conditions, for representing realistic dynamics, and for simulating heterogeneous chemical systems of multiple components. These many-body functional forms are typically more computationally expensive (3 to 10 times as expensive as the simplest molecular mechanics force fields), making HECC even more necessary. While the most recent generation of polarizable empirical energy functions provides some significant improvement, important work remains in how to model the physics of charge transfer between separate molecules and how to describe polarization anisotropy in fluctuating charge models. There are also algorithmic issues to be addressed to achieve computational efficiency in Monte Carlo simulations and extensions to arbitrary molecular systems.

However, the most fundamental expense in evaluating empirical force-field energies and derivatives is due to the long-range coulombic forces. The accounting of long-range forces is best introduced through the Ewald summation. Typical materials simulations periodically replicate the system in three spatial dimensions, and this approach divides the long-range coulombic interactions into a short-range part that is evaluated in real space (as a direct sum over atomic positions) and a long-range part evaluated in reciprocal space. New formulations of Ewald algorithms scale as $N \log N$ once N exceeds about 1,000, so systems with tens of thousands of atoms may reasonably be handled on the most advanced supercomputers.

Looking to the future, the new ab initio molecular dynamics approaches allow calculation of electronic structure on the fly, currently to an accuracy competitive with that of HF or DFT. Even though these methods are in their infancy and are not feasible for the long timescales and large molecular sizes that are needed for useful empirical force-field calculations, this capability will continue to grow over the next several decades. The bottlenecks for this area are primarily model physics (greater accuracy than that provided by HF and DFT), improved algorithms, and deployment on massively parallel architectures.

Improved Statistical Sampling

Phase equilibria calculations involve direct evaluations of thermodynamic properties of the individual phases at a series of state points to find where the temperature, pressure, and chemical potential of the phases are equal. Because the direct computation of conformational and energetic properties is so computationally expensive, as explained above, it is critical to be as efficient as possible in sampling the state points. Addressing that need, extended system equations of motion and associated numerical integrators have been developed that allow extensions from microcanonical ensemble dynamics to sampling of states in the canonical ensemble (NVT) as well as the isobaric-isothermal (NPT) ensembles. Based on a factorization of the evolution operator, a formal decomposition of the integration time step allows bonds to be updated more often than angle bends, and angle bends more often than short-range forces, and short-range forces more often than long-range forces. This formally correct multiple-time-step integration has been shown to generate about an order of magnitude improvement in computational efficiency in materials systems, although resonance artifacts can reduce this efficiency gain in practice. The decrease in computer time results from the fact that the most expensive terms, the double sum over atoms, need to be updated less often than local interactions. Calculations performed using multiple-time-step integration methods in isothermal or isobaric-isothermal ensembles are very scalable. Each time step results in a collective "move," and parallelization can proceed using standard domain decomposition paradigms. These are important considerations for phase equilibria calculations since large systems are required to overcome finite-size effects and heterogeneous systems require significantly longer equilibration times.

Probably the biggest breakthroughs in calculation of vapor-liquid and liquid-liquid phase equilibria are the formulations of grand canonical Monte Carlo methods in terms of the Gibbs ensemble and semi-grand canonical Monte Carlo methods developed in the 1980s and 1990s, allowing the determination of phase equilibria in one simulation without the interference of an imposed phase interface. Furthermore, once a single point on the coexistence curve is known, the rest of the curve can be calculated without resort to additional free-energy calculations by integrating the Clausius-Clapeyron equation, although care must be taken to avoid numerical instabilities. For solid-liquid phase equilibria, thermodynamic integration based on paths with a known free energy reference state are well developed. Enhanced sampling schemes and related methods are available that allow for efficient molecule exchanges between phases in order to converge the Gibbs ensemble.

Progress on Major Challenge 3 involves a very different focus, namely mathematical optimization research as a general approach for obtaining solutions to large nonlinear systems with numerous local minima. Constrained optimization methods rely on the availability of sufficiently well-defined constraints (supplied by the application expert) so that the desired solution is the only available minimum, or one of few available minima, in the optimization phase of the algorithm. Alternatively, global optimization techniques attempt to systematically search the parameter space based on a cost function to find all low-lying minima, including the global energy minimum. The useful application of these optimization strategies is computationally intensive since they typically require hundreds or thousands of evaluations of a cost function (and of its derivative if available). These optimization approaches are useful in many contexts, including atomic-level structure optimization of molecules in thermal-based separations and MSA materials in Major Challenges 1 and 2 and designs for entire separation systems that define Major Challenge 3. What distinguishes the usefulness of mathematical optimization in Major Challenges 1 and 2 is that the cost functions and parameter space are relatively well defined in terms of an objective function, while Major Challenge 3 has greater uncertainty in the nature and dimension size of the mathematical model.

In summary, our current capabilities in computational chemistry for addressing Major Challenges 1 and 2 are sufficient for a class of materials and chemical systems important for some separations problems, and improvements in HECC, broadly defined, will surely and steadily enable additional advances for these application areas, especially for larger molecular systems. However, advancing these methods to new classes of materials will require a combination of new model physics, better-scaling algorithms in quantum chemistry and statistical mechanical sampling, and deployment onto massively parallel architectures. Major Challenge 3 currently is more narrowly focused on formulating cost function models that can utilize the large array of mathematical optimization techniques.

OTHER ISSUES THAT LIMIT THE VALUE OF HECC TO CHEMICAL SEPARATIONS

Productive cooperation and dialogue between experimentalists and modelers is not as extensive as it should be in order for computational approaches to contribute optimally to progress in chemical separations. In particular, funding is lacking for experimental work to learn about phase equilibria in fundamental systems, knowledge that could be used to validate computational models. A combined computational/experimental strategy is critical. A good example of cooperation is the Industrial Fluid Simulation Challenge, sponsored by the National Institute of Standards and Technology (NIST), in which academic and industrial teams attempt to predict a range of thermodynamic and physical properties like vapor-liquid equilibria, density, viscosity, vapor pressure, heat of mixing, and so on. However, after four NIST Challenges, it is clear that there is a long way to go before computation can reliably predict various properties for a disparate set of chemical species.

Incentives are needed to encourage collaboration between experimentalists and researchers performing molecular simulations in order that computational models can be developed, validated, and run more efficiently. Within the research community in general, not much effort is being put into validating the results of atomistic scale models with experimental data. Part of the problem is that experimentalists have incentives to pursue experiments that are project-specific rather than those that will expand fundamental knowledge. Indeed, the measurements that would be most helpful in developing and verifying computational methods are often perceived as having little practical value for the experimentalist or funding agency. Cooperation between industry, academia, and government could create the needed incentives.

As noted in the preceding section, education and training are important if the chemical separations field is to profit from the potential of HECC. Developing and applying advanced simulation capabilities requires specialized cross-disciplinary skills; this topic is addressed in more detail in Chapter 7 because it affects nearly every field that relies on computational science and engineering. In the case of chemical separations in particular, these broad computational skills must be supported on the foundations of theoretical chemistry and mathematics and are vital to overcoming all three challenges explored in this chapter.

REFERENCES

Cho, S.H., B.Q. Ma, S.T. Nguyen, J.T., Hupp, and T.E. Albrecht-Schmitt. 2006. A metal-organic framework material that functions as an enantioselective catalyst for olefin epoxidation. *Chemical Communications* 24: 2563.

Department of Energy. 2005. *Hybrid Separations/Distillation Technology: Research Opportunities for Energy and Emissions Reduction*. DOE Industrial Technologies Program, Washington, D.C., April.

Dinca M., A.F. Yu, and J.R. Long. 2006. Microporous metal-organic frameworks incorporating 1,4-benzeneditetrazolate: Syntheses, structures, and hydrogen storage properties. *Journal of the American Chemical Society* 128: 8904.

Férey, G., C. Mellot-Draznieks, C. Serre, F. Millange, J. Dutour, S. Surblé, and I. Margiolaki. 2005. A chromium terephthalate-based solid with unusually large pore volumes and surface area. *Science* 309 (5743): 2040-2042.

Humphrey, J.L., and G.E. Keller. 1997. *Separation Process Technology*, New York, N.Y.: McGraw-Hill.

James, S.L. 2003. Metal-organic frameworks. *Chemical Society Reviews* 32: 276.

Jorgensen, W.L. 2004. The many roles of computation in drug discovery. *Science* 303 (5665): 1813-1818.

Kitagawa, S., R. Kitaura, and S.Noro. 2004. Functional porous coordination polymers. *Angewandte Chemie International Edition* 43: 2334.

Latroche, M., S. Surble, C. Serre, C. Mellot-Draznieks, P.L. Llewellyn, J.H. Lee, J.S. Chang, S.H. Jhung, and G. Ferey. 2006. Hydrogen storage in the giant-pore metal-organic frameworks MIL-100 and MIL-101. *Angewandte Chemie International Edition* 45: 8227.

Lipkowitz, K.B. 1998. Applications of computational chemistry to the study of cyclodextrins. *Chemical Reviews* 98: 1829-1873.

Martin, Y.C., M.G. Bures, E.A. Danaher, J. Delazzer, I. Lico, and P.A. Pavlik. 1993. A fast new approach to pharmacophore mapping and its application to dopaminergic and benzodiazepine agonists. *Journal of Computer-Aided Molecular Design* 7 (1) :83-102.

Matsuda, R., R. Kitaura, S. Kitagawa, Y. Kubota, R.V. Belosludov, T.C. Kobayashi, H. Sakamoto, T. Chiba, M. Takata, Y. Kawazoe, and Y. Mita. 2005. Highly controlled acetylene accommodation in a metal-organic microporous material. *Nature* 436 :238.

Millward, A.R., and O.M. Yaghi. 2005. Metal-organic frameworks with exceptionally high capacity for storage of carbon dioxide at room temperature. *Journal of the American Chemical Society* 127: 17998.

Mueller, U., M. Schubert, F. Teich, H. Puetter, K. Schierle-Arndt, and J. Pastré. 2006. Metal-organic frameworks—prospective industrial applications. *Journal of Materials Chemistry* 16: 626.

NRC (National Research Council). 1998. *Separation Technologies for the Industries of the Future*. Washington, D.C.: National Academy Press.

NRC. 2000. *Alternatives for High-Level Waste Salt Processing and the Savannah River Site*. Washington, D.C.: National Academy Press.

Pan, L., D.H. Olson, L.R. Ciemnolonski, R. Heddy, and J. Li. 2006. Separation of hydrocarbons with a microporous metal-organic framework. *Angewandte Chemie International Edition* 45: 616.

Panella, B., M. Hirscher, H. Pütter, and U. Müller. 2006. Hydrogen adsorption in metal-organic frameworks: Cu-MOFs and Zn-MOFs compared. *Advanced Functional Materials* 16: 520.

Rablen, P.R., J.W. Lockman, and W.L. Jorgensen. 1998. Ab initio study of hydrogen-bonded complexes of small organic molecules with water. *Journal of Physical Chemistry A* 102: 3782-3797.

Rosseinsky, M.J. 2004. Recent developments in metal-organic framework chemistry: Design, discovery, permanent porosity and flexibility. *Microporous Mesoporous Materials* 73: 15.

Snurr, R.Q., J.T. Hupp, and S.T. Nguyen. 2004. Prospects for nanoporous metal-organic materials in advanced separations processes. *American Institute of Chemical Engineers Journal* 50: 1090.

Walters, W.P., M.T. Stahl, and M.A. Murcko. 1998. Virtual screening—An overview. *Drug Discovery Today* 3(4): 160-178.

Wang, Q.M., D. Shen, M. Bülow, M.L. Lau, S. Deng, F.R. Fitch, N.O. Lemcoff, and J. Semanscin. 2002. Metallo-organic molecular sieve for gas separation and purification. *Microporous Mesoporous Materials* 55: 217.

Westmoreland, P.R., P.A. Kollman, A.M. Chaka, P.T. Cummings, K. Morokuma, M. Neurock, E.B. Stechel, and P. Vashishta. 2002. *Applications of Molecular and Materials Modeling*. World Technology Evaluation Center Panel Report. Available at http://www.wtec.org/loyola/molmodel/mm_final.pdf.

6

Numerical and Algorithmic Characteristics of HECC That Will Be Required by the Selected Fields

This chapter discusses the numerical and algorithmic characteristics of HECC that will be needed in the four fields examined in Chapters 2-5. In particular, it addresses tasks (e) and (f) of the charge: "Identify the numerical and algorithmic characteristics of the high-end capability computing requirements needed to address the scientific questions and technological problems identified in [Chapters 2-5]" and "Categorize the numerical and algorithmic characteristics, specifically noting those categories that cut across disciplines." Among other things, this chapter gives some indication of the mathematics, computer science, and computing infrastructure requirements, opportunities, and difficulties associated with the opportunities identified in Chapters 2-5. It identifies the prevailing computational demands in each of the four fields and suggests how the demands are likely to grow in the near and longer term. It also characterizes the rate-limiting mathematical parts of these calculations.

In this chapter the committee discusses two stimulating tasks facing the four fields of science and engineering covered in this report: (1) managing and exploiting massive (and growing) amounts of data and (2) preparing the next generation of people who will push the frontiers of computational science and engineering. Both tasks are pervasive across many other fields of science and engineering as well.

NUMERICAL AND ALGORITHMIC CHARACTERISTICS OF HECC FOR ASTROPHYSICS

Several tasks face HECC for astrophysics in the near term. One is to raise the overall performance of astrophysics codes to the point where the current generation of HECC platforms can be effectively used. At present, only a small fraction of the algorithms and codes in use scale to 10^3-10^4 processors. Some algorithms—for example, grid-based fluid dynamics—scale very well to tens of thousands of processors, while others require global communication that can limit scaling—for example, elliptic partial differential equations, such as Poisson's equation. The availability of systems with 10^5 or more processors will enable these much larger calculations while also making it feasible to perform more complex calculations that couple different models. Algorithms, models, and software are needed to enhance scalability, especially for those computations that require adaptive mesh refinement (AMR) or multiscale methods. The scalability of software is limited by the performance of its least-scalable component. As we incorporate more models into one code (multiscale and/or multiphysics models) and perhaps integrate data management and other functions, it is more likely that overall performance will be held back by one weak link. Therefore, there will be an increased need to work toward balanced systems with components that are relatively similar in their parallelizability and scalability.

Another problem is that the computations must deal with an enormous number of particles (10^{10}-10^{11} at present) and large grids (as big as 2048^3, or 10^{10} cells, currently, with larger calculations planned for the future). Complex methods can require 10^3-10^4 floating point operations per cell per time step and generate hundreds of gigabytes to multiple terabytes of data in a single snapshot. Systems that deal with data on this scale will usually need to access thousands, perhaps even hundreds of thousands, of disks in parallel to keep the system's performance form being limited by input/output (I/O) rates. Thus scalable parallel I/O software would be extremely critical in these situations in the long term.

In all of the HECC-dependent areas of astrophysics reviewed in Chapter 2, there is a strong science case to use the increase in computational capability expected over the next 10 years to increase model fidelity. This would be done mainly by including the effects of a large number of physical processes and their interactions in a single simulation. For such a program to be successful, it will be necessary to make fundamental improvements in the models, algorithms, and software.

Models

Traditionally, large-scale simulation in astrophysics has taken the form of first-principles modeling, in which the equations to be solved were relatively well characterized. Examples include three-dimensional calculations of inviscid compressible flow, flows of collisionless matter, effects of self-gravity, and passive radiative losses in optically thin media. The next generation of models will go in the direction of much more complicated coupling between different physical processes (multiphysics modeling) and the interaction of a much greater range of temporal and spatial scales (multiscale modeling). Some of these complex models include the following:

- In cosmology models, those that incorporate the feedback effects of star formation and of supermassive black holes on galaxy formation.
- In models of the early stages of star formation, those that incorporate a self-gravitating multicomponent medium that includes both collisional and collisionless components undergoing ionization, chemical reactions, heat and mass transfer within and among the components, and radiative transfer.
- In models of supernovae, those that include the ignition and propagation of a nuclear reaction front in a turbulent, gravitationally stratified medium, on the scale of an entire star.

In these and other cases, it is not feasible to use a first-principles model that resolves all of the length scales and timescales and all of the coupled physical processes, even under the most optimistic view of the growth of computational capabilities over the next decade. Instead, it will be necessary to resolve a narrower range of length scales, timescales, and physical processes, with the remaining effects represented approximately through models that can be represented on the resolved scales. The development of such models is a complex process involving a combination of mathematical analysis and physical reasoning; they must satisfy the mathematical requirement of well-posedness in order to be computable; and they must be validated. Large-scale, detailed simulations will play a substantial role throughout this process by simulating unresolved scales that will provide input to the development of models and be the basis for validation.

Algorithms

For the simulation of large-scale phenomena, such as cosmology, star formation, or gravity wave formation and propagation, the use of multiresolution for particles (hierarchical N-body methods) or fields (AMR methods) are core capabilities on which much of the current success is built. However, multiresolution methods are mature to varying degrees, depending on the level of model complexity. AMR for magnetohydrodynamics or for Einstein's equations of general relativity, for example, is currently undergoing rapid development, while for coupled radiation and matter, or general relativistic fluid dynamics, such methods are still in their infancy. Radiation is particularly difficult owing to the need to solve time-dependent problems in six-dimensional phase space. For supernova simulations, an additional set of difficulties exists. For instance, while for much of the simulation the star is spherically symmetric on the largest scales, it has asymmetric three-dimensional motions on the small scales. The need to preserve that large-scale symmetry requires new gridding methodologies such as moving multiblock grids, combined with local refinement. A second difficulty is that there are stiffness issues due to the reaction kinetics of thermonuclear burning or due to low-Mach-number fluid flows. New algorithms will need to be developed to integrate over the fast timescales efficiently and without loss of accuracy or robustness.

Data Analysis and Management

For data-intensive fields like astronomy and astrophysics, the potential impact of HECC is felt not just in the power it can provide for simulations but also in the capabilities it provides for managing and making sense of data, irrespective of whether the data are generated by simulations or collected via observations. The amount, complexity, and rate of generation of scientific data are all increasing exponentially. Major Challenges 1 and 2 in Chapter 2 (the nature of dark matter and dark energy) probably will be addressed most productively through observation. But whether data are collected through observation or generated with a computer, managing and exploiting data sets on this scale is critically dependent on HECC. The specific problems stemming from massive amounts of data are discussed in Chapter 2.

Software Infrastructure

There are three drivers for the development of software infrastructure for astrophysics in the long term. The first is the expected radical change in computer hardware. Gains in aggregate performance are expected to come mainly from increasing the level of concurrency rather than from a balanced combination of increases in clock speeds of the constituent processors and increases in the number of processors. Only a small fraction of the algorithms of importance to astrophysics have been shown to scale well to 10^3-10^4 processors. Thus high-end systems requiring the effective use of 10^8 processors and 10^9 threads represent an enormous challenge. The second driver is the nonincremental nature of the model and algorithm changes described above. Development of optimal codes will require an aggressive and nimble exploration of the design space. This exploration will involve a moving target because it will have to be done on very high-end systems, whose architectures are evolving simultaneously. The third driver is the problematic aspects of data management.

The response to these three drivers is roughly the same: Design high-level software tools that hide the low-level details of the problem from the developer and user, without foreclosing important design options. Such tools will include new programming environments to replace MPI/OpenMP for dealing with new hardware architectures. For new algorithm development, the tools are software frameworks and libraries of high-level parallel algorithmic components (for example, discretization libraries, data holders, solvers) from which new simulation capabilities can be built. Similar collections of visualization, data-analysis, and data-management components would provide capabilities for those tasks. Some methods and prototypes have already been developed by the mathematics and computer science research communities that would form a starting point for developing these high-level software tools. They would have to be customized in collaboration with the astrophysics community.

NUMERICAL AND ALGORITHMIC CHARACTERISTICS OF HECC FOR THE ATMOSPHERIC SCIENCES

In the atmospheric sciences, the requirements, opportunities, and challenges divide roughly into near term (1-5 years) and long term (5-10 years). In the near term, the main opportunities require the use of existing simulation capabilities on 10^3-10^4 processors to address scientific questions of immediate interest. In the longer term, the ability to exploit capabilities that use 10^4-10^5 or more processors per run on a routine basis will require the development of new ideas in mathematical models, numerical algorithms, and software infrastructure. The core computations are those associated with computational fluid dynamics, though many other models and algorithms—for example, gridding schemes, statistical models of subgrid-scale processes, models of chemical kinetics, statistical sampling across ensembles, and data-management models—are also essential.

Near-Term Requirements, Opportunities, and Challenges

Currently, simulation codes for climate and numerical weather prediction (NWP) are routinely run on 700-1,500 processors. Nevertheless, both efforts are still constrained by the insufficient aggregate computing capacity available to them. In the case of climate, the atmospheric models are run at horizontal mesh spacing of about 100 km, which is insufficient to capture a number of key solution phenomena, such as orographically driven precipitation. To resolve such phenomena, it is necessary to increase the horizontal mesh resolution by a factor of four (25 km mesh spacing). In the case of NWP, the demand for increased capability is driven by requirements for improved prediction of severe weather and for better support of critical industries such as transportation, energy, and agriculture. In addition, new higher-resolution data streams are expected to soon be available for NWP, and the computational effort involved in data assimilation will increase accordingly. Thus, for both climate modeling and NWP, there is a near-term requirement to increase computing capability by a factor of 10 or so. In the case of climate modeling, this increased capability could be used to increase the atmospheric grid resolution of a coupled atmosphere-ocean-sea ice climate model. Such simulation codes have already been demonstrated to scale up to as many as 7,700 processors, so no reimplementation would be required. In the case of NWP, the increase of computer power would need to be accompanied by changes in code architecture to improve the scalability to 10^4 processors and by the recalibration of model physics and overall forecast performance in response to the increased spatial resolution.

Long-Term Requirements, Opportunities, and Challenges

A number of opportunities will open up as computing at the petascale and beyond becomes available. For climate simulations, these include the prediction of global and regional precipitation patterns with an accuracy and robustness comparable to what we now have for temperature; the accurate prediction of effects that would have substantial human impacts, such as climate extremes; and decadal predictions on regional scales, which are of greatest interest currently to policy makers. In NWP, such models would enable "warn on forecast" for locally severe weather, improved prediction of hurricane intensity and landfall, and very high-resolution analysis products. In order to realize these opportunities, there will need to be major advances in models, algorithms, and software.

Models

For both climate modeling and NWP, one of the most prominent changes in the models that will enable the advances outlined above is the replacement of the current statistical models of convective and hydrological processes by much higher-fidelity explicit representations. This change is intimately tied to increased spatial resolution (horizontal mesh spacing of ~1 km), which we will discuss below. However, development of the models themselves is a substantial mathematical undertaking. For convective processes, even at 1 km resolution, there are smaller-scale motions that are not representable on the grid, for which large-eddy simulation models will need to be developed. For hydrological processes, the problem is even more complicated, with clouds having a geometrically and thermodynamically rich microphysical structure whose effects must be scaled up to the scale of cloud systems. Finally, for climate modeling, the need for high-fidelity representations of clouds in terms of their hydrology and their interaction with radiation introduces additional difficulty. Addressing these problems will require a combination of mathematical and computational techniques to produce simulations at scales small enough to explicitly model the convective and hydrological processes of interest. That effort must be strongly coupled to observations and experiments so that the simulations can be validated and the subgrid-scale models developed from those calculations can be constrained by experimental and observational data. This process will incorporate new ideas being developed in the multiscale mathematics community that will provide techniques for using such small-scale simulations to inform models at the larger scales.

Algorithms

One of the principal drivers of algorithm development in atmospheric models is the need for substantial increases in spatial resolution, both in climate modeling and NWP. There are two possible approaches to obtaining resolutions of a few kilometers. The first is to simply use a uniform grid at that resolution. Such a discretization of an atmospheric fluid dynamics model would strain or exceed the limits of a petascale system. A second approach is to use a multiresolution discretization, such as nested refinement, provided that the regions that require the finest resolution are a small fraction (10 percent or less) of the entire domain. In that case, the computational capability required could be reduced by an order of magnitude, which would make the goal of computing with such ultra high-resolution models more feasible. As shown by Figure 3-4, such multiresolution methods are already feasible in weather prediction; however, they are not yet in use in most climate models. In any case, a broad range of design issues would need to be addressed before such models could be used routinely in climate or NWP, including the choice of discretization methods, coupling between grids at different resolutions, and dependence of subgrid models on grid resolution.

In current atmospheric simulation codes, there is the additional difficulty that the time step must decrease in proportion to the spatial mesh size in order to achieve numerical stability. This further increases the computing power needed for finer resolution and limits the potential gain from simply applying more powerful computers. This is a particularly pressing problem for climate modeling, because it calls for evolving the system for very long times. One possible solution would be to treat more of the time evolution implicitly, so that larger time steps can be taken. There have been many new ideas for designing implicit methods, such as matrix-free Newton-Krylov methods for building efficient solvers for the resulting systems of equations and deferred corrections methods for designing implicit and semi-implicit time discretizations.

A second class of algorithms that is limiting current capabilities in atmospheric science contains methods for data assimilation and management. While data assimilation techniques are a mature technology in atmospheric modeling, in climate modeling and NWP new problems are arising that call for a reconsideration of data assimilation. In climate modeling, new data on biogeochemistry and the carbon cycle are being collected for the first time. Both the nature of the data and the nature of the models will require a rethinking of how to integrate these data into the overall assimilation process for climate modeling. In the case of NWP, growth in the volume of data is the dominant change, and the required four-dimensional assimilation will require new, more efficient methods for managing, analyzing, mining, sharing, and querying data and considerably more computing power.

Software Infrastructure

The comments about software infrastructure for the long-term needs of astrophysics apply equally well to the atmospheric sciences. But the atmospheric sciences face some additional issues because some of their computational products are used operationally for NWP. For that reason, changes in algorithms will have to be evaluated at the high spatial and temporal resolutions that we expect to use in next-generation operational systems. Higher-resolution validation—a common requirement in computational science and engineering—demands even more computing capability than operational predictions. Another difference between computing for the atmospheric sciences and for astrophysics is that data management decisions for the atmospheric sciences must take into account the needs of a much broader range of stakeholders than just the scientific community. Finally, any new methods and prototype software developed in the atmospheric sciences must be made robust enough to produce forecasts reliably on a fixed production schedule.

NUMERICAL AND ALGORITHMIC CHARACTERISTICS OF HECC FOR EVOLUTIONARY BIOLOGY

Computations of importance for evolutionary biology rely heavily on statistics and methods from discrete mathematics, such as tree searching, data mining, and pattern matching. The computational methods for searching and comparing genomes are increasingly important. A particular challenge is generating and validating phylogenetic trees as they grow to include thousands of species. Algorithms for all such computations tend to scale poorly. However, that has not been a significant limitation to date because computational evolutionary biology is still young enough that important scientific results can be obtained by the study of modest systems. Such systems can be studied in many cases with desktop computing, and that is the prevailing scale of computing today in evolutionary biology, although there are exceptions. Working with one's own desktop system has clear advantages when a field is still exploring a diversity of models, algorithms, and software, because an investigator can customize and adjust the computing environment in many ways.

However, this era is ending. As the community gains confidence in particular models, as algorithms are improved, and as data are assembled, it is inevitable that researchers will strive to study larger and larger systems. Adding more species and making more use of genomic data will quickly drive computational evolutionary biology into the realm of high-end computing. Because of scalability problems with many algorithms and the massive amounts of genomic data to be exploited, evolutionary biology will soon be limited by the insufficiency of HECC.

In the longer term—once the foundations for simulation as a mode of inquiry become well established—the numerical, algorithmic, and other related infrastructure requirements of evolutionary biology will be very similar to those of astrophysics and the atmospheric sciences. A particularly exciting longer-term impact of the use of HECC for evolutionary biology is that the dynamic interplay of ecological genetics, evolutionary genetics, and population genetics will be studied, whereas today ecological theory and evolutionary theory have most often been studied in isolation, as noted in Chapter 4. As noted above, evolutionary biologists are beginning to add population details. Access to high-end computation will enable increased coupling despite the different timescales of ecological and evolutionary processes. Applying HECC will allow making the models consistent with life processes, which couple ecological and evolutionary dynamics. Thus the impact will begin near term but extend into the indefinite future as this very important question, the evolutionary dynamics of the phenotype-environmental interface, is addressed.

Another powerful capability resulting from HECC will be the introduction of climate models into evolutionary trees, which will connect environmental and ecological modeling with evolutionary modeling. This would provide the basis for, among other topics, causal models that link Earth's physical history (including climate change) with its biotic evolution, necessitating extraordinary spatiotemporal scales. Models that link Earth's geosphere and its biosphere will inevitably have a large number of parameters. Already, scientists across many disciplines couple climate and environmental models as well as ecosystem distribution (from satellite data) to simulate the course of Earth's environment and predict changes due to global change. Species distribution can be modeled as a function of environment or environmental models, which can lead to inferences about the environmental experiences of common ancestors down a phylogenetic tree. Thus these classes of environmental and evolutionary models can be linked and extended over historical time to look at the interplay of the physical world (geology, climate) and the biological world (species, populations, communities within the range of natural environments). The powerful advances beginning today will mature, along with related efforts to integrate and analyze disparate data sets. This will lead to very large requirements at this interface that will also push whatever state-of-the-art computing resources and infrastructure are available in the 5- to 15-year time frame.

Besides the near-term opportunities for high-end computing to provide insight into speciation, the continued access to advanced resources will open up new options in the longer term for refined analysis of the origin of individual species and the speciation process. The option of including demographic history will become viable. Sequence data yield multiple phylogenies, and complex population histories are often consistent with a number of different sequence trees, resulting in considerable uncertainty. Nor can analytical or closed-form solutions be realized. Approximations will continue to be used in the short term, including Markov chain Monte Carlo sampling. Increased computing power and the continued analysis of speciation will enable the statistical evaluation of the universe of potential trees, and then population models will need to be evaluated to see the best fit to the trees. In sum, as theory advances and allows the exploration of alternatives, continued progress on the speciation problem will require access to whatever state-of-the-art capability platform and associated computing infrastructure become available.

At the heart of what biologists term twenty-first century biology is the characterization of the truly complex processes of multicellular organisms and their development. Today, this work is just beginning: An experimental basis, with some theoretical foundation, is being established. As always, the actual characterization must proceed from an evolutionary biology perspective. Understanding the evolution of cell processes and of development—how intracellular and intercellular (physiological) networks and cellular- to organismic-level developmental processes have evolved—will require implementing new algorithms as well as continuing the current, rapid experimental progress. These efforts are just beginning as biologists learn that simple extensions of engineering and physical models are wholly inadequate for capturing the complexity of any level of biology and life's processes. In the next 5-15 years, massive computational efforts will be involved in such modeling.

NUMERICAL AND ALGORITHMIC CHARACTERISTICS OF HECC FOR CHEMICAL SEPARATIONS

The key HECC requirements of chemical separations are those that will allow simulations of greater complexity to be performed more accurately so that their ability to guide experimentation can be exploited more readily. Current algorithmic formulations provide very useful physical insights, but they are not yet sufficiently accurate to provide stand-alone predictive power for phase-equilibria problems nor powerful enough to represent the complexity needed for separation process design. At the least they require a close interface with experimental validation studies. However, experimental approaches typically are limited to optimizing existing chemical separations solutions. Computational approaches, by contrast, offer low-cost explorations of radical changes that might optimize chemical separations.

While current model and algorithmic capabilities allow the routine use of 10^3-10^4 processors per run, long-term requirements necessary for exploiting 10^5-10^6 or more processors will require the development of new ideas in mathematical models, numerical algorithms, software infrastructure, and education and training in the computational and mathematical sciences.

Given that some 80 percent of the chemical separations industry essentially relies on understanding phase equilibria—whether explicitly formulated as a thermal-based distillation problem or as the context for developing MSA materials—the current capabilities of computational chemistry must be significantly extended to address Major Challenges 1 and 2 in Chapter 5. Computational chemistry includes calculations at the molecular scale with algorithms based on quantum chemical theories and classical analogs that evaluate the energetics of molecular conformations, as well as statistical mechanical methods that sample those conformations consistent with thermodynamic variables such as temperature and pressure. The underlying theoretical framework of quantum mechanics is used to define a model potential energy surface for the materials system of interest, and statistical mechanics is used to formulate the sampling protocols on this surface to evaluate equilibrium properties at a series of thermodynamic state points.

Simulation methods such as molecular dynamics involve calculating averaged properties from finite-length trajectories. The underlying molecular dynamics engine is a particle-based algorithm that solves Taylor expansion approximations to Newton's equation of motion. The algorithms are well-formulated as symplectic integrators—that is, discretizations that have a Hamiltonian structure analogous to that of the dynamics being approximated, which contributes to their ability to generate stable long trajectories. Molecular dynamics simulations involve two levels of problem granularity, which makes them well-suited for parallelism. The rate-limiting step for these simulations is the evaluation of empirical energy and forces for N particles. The most common forms of those energies and forces map onto a fine-grained parallelization (using spatial or force decomposition) that scales reasonably well (approximately 55 percent of linear) up to 128 processors. More sophisticated models include multibody polarization, or multipoles, which scale less well and will require reconsideration of problem granularity.

Overlaid on this fine-grained parallelization is another layer of (trivial) coarse-grained parallelization involving a statistical mechanical sampling algorithm which runs N independent simulations, which may involve infrequent communication to swap state point information ($6N$ real numbers). Both MPI and

OpenMP are already effectively used to efficiently exploit distributed and shared memory architectures. Calculations involving 10^5 particles are currently feasible with the use of existing simulation capabilities on 10^3-10^4 processors. In addition, improved sampling methods that accelerate the convergence to equilibrium, characterize dynamical properties involving long timescales, and sample rare events have been advanced by impressive new mathematical models such as transition path theory formulations and the string method and its variants. The numerical algorithms are well understood; coupled with longer trajectory runs on larger systems, such algorithms will easily be deployed on massively parallel architectures involving 10^5-10^7 processors.

Currently, quantum electronic structure algorithms are used to develop empirical force fields by calculating conformational energies and geometries of small material fragments that are ultimately transferred to describe components of larger materials. For some materials, the physics of electron correlation is well-described by a mean field theory plus a second-order perturbation of electron correlations (MP2). MP2 is formulated as an algorithm that solves dense linear algebra equations dominated by matrix multiples. Significant algorithmic improvements over the last 5 years allow for MP2 calculations of 40-50 atom systems on a single processor; recent modestly scaling parallel algorithms allow for MP2 calculations on larger chemical systems of up to a few hundred atoms. Computational hardware improvements would allow for the MP2 calculation of much larger fragments and model compounds, which would reduce the error in transforming the quantum mechanical data into parameters for empirical, classical potential energy surfaces of the chemical separations materials of interest.

But for other materials, MP2 is inadequate or fails altogether. To attain the desired quantum model accuracy in those cases requires the use of more sophisticated electron correlation schemes for wave-function methods or developing the promise of density functional theory (DFT). Higher-order correlation schemes for wave-function methods are well-formulated mathematical models, but they are severely limited by poor algorithmic scaling. By contrast, DFT methods scale much better but are currently limited by weaknesses in their theoretical formulation. Future capabilities in computing and algorithms would allow the use of the gold standard, coupled cluster algorithms. Because these are not yet parallelized, applying them to fragments with more than 10 atoms is not yet feasible in the absence of symmetry.

Major Challenge 3 concerns the design of overall separation systems. It is an instance of a mathematical optimization problem that aims to extremize a relevant cost function. For example, methods have been developed that use mathematical "superstructure optimization" to find optimal configurations and operating conditions for separations based on inputs about mass, heat, and momentum transfer. However, superstructure optimization is limited by the difficulty of formulating the full parameter space for the initial superstructure and the large size of the optimization problem necessary to determine the best solution within that space. The first of these limitations requires expert input to help set up the parameter space of the problem. The second can be addressed with new optimization approaches known as generalized disjunctive programming that can deal effectively with the discontinuous and nonconvex nature of the resulting mixed-integer nonlinear program. But such mathematical optimization problems quickly push the limits of HECC as more complexity is incorporated into the model.

In summary, computational chemistry is adequate at present to address, usually only qualitatively, the most standard chemical separations processes of immediate interest to industry, although it could also serve to screen candidate processes for nonstandard phase-equilibria chemical separations. The primary benefit of future massively parallel computers would be to increase by one or two orders of magnitude the size of chemical systems that can be simulated, and sampling timescales can be significantly lengthened, leading to better convergence of phase data without any significant changes in algorithmic structure. Enhanced hardware capabilities will allow superstructure optimization to be applied to usefully address complex synthesis problems. That having been said, far more issues remain—model accuracy (computational chemistry), model development (superstructure parameter space), and workforce training in computational and mathematical sciences—to enable moving ahead on the biggest challenges facing chemical separations.

CATEGORIZATION OF NUMERICAL AND ALGORITHMIC CHARACTERISTICS OF HECC NEEDED IN THE FOUR SELECTED FIELDS

Models

A common thread emerging from this study of four fields is the need to develop models whose detailed mathematical structure is still not completely specified, much less understood. In astrophysics and the atmospheric sciences, the need for new models arises from the attempt to represent complex combinations of physical processes and the effect of multiple scales in a setting where many of the constituent processes (fluid dynamics, radiation, particle dynamics) have well-defined mathematical descriptions whose ranges of validity are well understood. In the field of evolutionary biology, comprehensive frameworks must be developed for mathematical modeling at all scales, from populations and ecosystems down to the cellular level, as well as connections to various levels of description and to experimental and observational data. In chemical separations, the complete parameter space for process optimization—that is, the space that contains the best solution—remains ill-defined and is developed in mixed discrete and continuous variables that do not allow for the ready use of off-the-shelf algorithms for mathematical optimization. In all four fields, the resulting models must be mathematically well-posed (in order for them to be reliably computable) and susceptible to validation (in order to obtain well-characterized ranges for the fidelity and applicability of the models).

Although the detailed requirements for models in these four fields are quite different, there are commonalities that suggest overlap in the mathematical infrastructure. One is the extent to which models will be bootstrapped from data. None of the new models will be first-principles models in the sense that some of the fundamental models in the physical sciences from the nineteenth and early twentieth centuries are. Those early models are broadly applicable and well-characterized mathematically and physically, and only a small number of parameters for them must be determined from experimental data. In contrast, the new models will probably be more domain-specific, with larger sets of parameters that will have to be determined from combinations of experimental and observational data and auxiliary computer simulations. This suggests tight coupling between model formulation and model validation, requiring optimization and sensitivity analysis. A second commonality is the extent to which the next generation of models is likely to consist of hybrids that have both deterministic and stochastic components. The well-posedness of such hybrids is much more difficult and far less complete than for pure deterministic or pure stochastic models, particularly when they are expressed in the form required to ensure well-behaved numerical simulations.

Algorithms

Several requirements occur repeatedly among applications in the physical sciences. One is the need for handling stiffness in time-dependent problems so that the time step can be set purely by accuracy considerations rather than by having to resolve the detailed dynamics of rapidly decaying processes as they come to equilibrium. In ordinary differential equations, this problem is dealt with using implicit methods, which lead to large linear systems of equations. Such an approach, when applied naively to the high-resolution spatial discretizations that arise in astrophysics and atmospheric modeling, leads to linear systems whose direct solution is prohibitively expensive in terms of central processing unit time and memory. The alternatives are to use reduced models that eliminate the fast scales analytically or to use iterative methods based on approximate inverses ("preconditioners") constructed to be computationally tractable and to take into account the analytic understanding of the fast scales. In both cases, a successful attack on this problem involves understanding the interplay between the mathematical structure of the slow and fast scales and the design of efficient solvers.

There is also a need for continuing advances in multiresolution and adaptive discretization methods, particularly in astrophysics and atmospheric modeling. Topics that require further development include the extension to complex multiphysics applications, the need for greater geometric flexibility, the development of higher-order accurate methods, and scalability to large numbers of processors. The construction of such methods, particularly in conjunction with the approaches to stiff timescales, described above, will require a deep understanding of the well-posedness of the problems being solved as initial-value or boundary-value problems in order to obtain matching conditions across spatial regions with different resolutions.

Finally, the need for high-performance particle methods arises in both astrophysics and chemical separations. The issues here are the development of accurate and efficient methods for evaluating long-range potentials that scale to large numbers of particles and processors and of stiff integration methods for large systems of particles.

Software

The two key issues that confront simulation in all four fields relate to the effective use of the increased computing capability expected over the coming decade. The first is that much of the increased capability will be used to substantially increase the complexity of simulation codes in order to achieve the fidelity required to carry out the science investigations described in Chapters 2-5.

The second issue is that much of the increase in capability is likely to come from disruptive changes in the computer architectures. Over the last 15 years, the increase in computing capability has come from increases in the number of processors, improved performance of the interconnect hardware, and improved performance of the component processors, following Moore's law, as clock speeds increased and feature sizes on chips decreased. As seen by the applications developer, these improvements were such that the computing environment remained relatively stable.

The difficulty we face now is that the continuous increase in single-processor performance from increased clock speeds is about to come to an end, because the power requirements to drive such systems are prohibitive. We expect that the aggregate performance of high-end systems will continue to increase at the same rate as before. However, the mechanisms for achieving that increase that are currently under discussion—hundreds of processors on a single chip, heterogeneous processors, hardware that can be reconfigured at run time—represent radical departures from current hardware and may well require an entirely new programming model and new algorithms. The potential rate of change here is overwhelmingly greater than what we have experienced over the last 15 years. Initially, HECC multiprocessor systems had ~100 processors, and current HECC systems have ~10^4 processors. By comparison, over the next decade we could see as many as 10^8 processors and 10^9 threads in order to see the same rate of increase in capability. It is not known how to effectively manage that degree of parallelism in scientific applications, but it is known that many of the methods used now do not scale to that level of concurrency.

The increased complexity of simulation codes that will be brought about by these two issues not only poses an economic problem but also could be a barrier to progress, greatly delaying the development of new capabilities. The development of simulation and modeling software, particularly for nonlinear problems, is a combination of mathematical algorithm design, application-specific reasoning, and numerical experimentation. These new sources of software complexity and uncertainty about the programming model threaten to stretch out the design-development-test cycle that is central to the process of developing new simulation capabilities.

Both issues suggest that it may be necessary to rethink the way HECC scientific applications are developed. Currently, there are two approaches to HECC code development. One is end-to-end development by a single developer or a small team of developers. Codes developed in this fashion are sometimes made freely available to the scientific community, but continuing support of the codes is not formally funded. The second approach is community codes, in which software representing complete simulation capabilities is used by a large scientific community. Such codes require a substantial investment for development, followed by funding for ongoing enhancement and support of the software. The single-developer approach is used to various extents by all four fields, with community codes used mainly in the atmospheric sciences and chemical separations. There are limits to both approaches relative to the issues described above. The single-developer approach is resource-limited: Building a state-of-the-art simulation code from scratch requires more effort and continuity than can be expected from the typical academic team of graduate students and postdoctoral researchers. The limitation of the community code approach as it is currently practiced is that the core capabilities of such codes are fixed for long periods of time, with a user able to make only limited modifications in some of the submodels. Such an approach may not be sufficiently nimble to allow for the experimentation in models and algorithms required to solve the most challenging problems.

An intermediate approach would be to develop libraries that implement a common set of core algorithms for a given science domain, out of which simulation capabilities for a variety of models could be assembled. If such a sufficiently capable collection of core algorithms could be identified, it would greatly expand the range of models and algorithmic approaches that could be explored for a given level of effort. It would also insulate the science application developer from changes in the programming model and the machine architecture, since the interfaces to the library software would remain fixed, with the library developers responsible for responding to changes in the programming environment. Such an approach has been successful in the LAPACK/ScaLAPACK family of libraries for dense linear algebra. Essential to the success of such an endeavor would be developing a consensus within a science domain on the appropriate core algorithms and the interfaces to those algorithms. The discussion above suggests that all four fields represented in this study would be amenable to such a consensus. Two of them already make use of community codes, and the other two have indicated the need for standard tool sets. More generally, the prospect of radical changes in all aspects of the HECC enterprise—hardware, programming models, algorithms, and applications—makes it essential that the computational science community, including both applications developers and library developers, work closely with the computer scientists who are designing the hardware and software for HECC systems.

CROSSCUTTING CHALLENGES FROM MASSIVE AMOUNTS OF DATA

Of the four fields examined in this study, three—astrophysics, atmospheric science, and evolutionary biology—can be characterized as very data intensive. The nature of their data, the types of processing, and the interaction between models and simulations vary across these three fields, with evolutionary biology being the most distinct.

The data intensity can be characterized using various dimensions:

- Size,
- Scaling,

- Complexity,
- Types of data,
- Processing requirements,
- Types of algorithms and models to discover knowledge from the data,
- Archiving information,
- Sharing patterns among scientists,
- Data movement requirements, and
- Impact of discovery and results in driving experiments, simulations, and control of instruments.

Data sets in all three of these fields, whether produced by simulations or gathered via instruments or experiments, are approaching the petascale and are likely to increase following an analog of Moore's law. First, new data from atmospheric and environmental observations are added every day as satellite and other sensor-based data are gathered at ever-increasing resolutions. This data resource is more than just a static archive: Continuing progress in scientific understanding and technological capability makes it possible to upgrade the quality and applicability of the stored data using analysis and simulations, which further increases the demands on storage and analysis. Second, as more instruments are used with higher resolution, the amount of observed data is continuously increasing at an exponential rate. Third, data produced by simulations are increasing as finer resolutions are used and larger simulations are performed.

The most challenging data-related aspects in these fields are the following:

- Discovery of knowledge from data in timely ways,
- Sharing and querying of data among scientists,
- Statistical analysis and mining of data, and
- Storage and high-performance I/O.

Advances in these fields require the development of scalable tools and algorithms that can handle all of these tasks for petascale data.

To improve the productivity of scientists and engineers, as well as that of systems, data management techniques need to be developed that facilitate the asking of questions against data and derived data, whether produced by simulations or observations, in such a way that simulations do not need to be repeated. Technologies that create these data repositories and provide scalable analytics and query tools on top of them are necessary, as is the development of ontologies and common definitions across these fields.

One of the most challenging needs for astrophysics and atmospheric science is the ability to share observational data and simulation data along with derived data sets and information. The atmospheric sciences are further along in providing common formats and sharing of data, but a tremendous amount of work remains to be done. One could think of this entire process as developing scientific data warehouses with analytical capabilities. Another very important aspect of data management and analysis in these domains is the development of paradigms and techniques that permit user-defined processing and algorithmic tasks no matter where the data reside, thereby avoiding or reducing the need for transporting raw tera- and petascale data. Acceleration hardware and software may both be useful here.

In evolutionary biology, the type, collection, and processing of data are somewhat different from what is encountered in astrophysics and atmospheric sciences and, in their own ways, very challenging. A large amount of data processing in evolutionary biology entails manipulating strings and complex tree structures. In the analysis of genomic data, the underlying algorithms involve complex searches, optimizations, mining, statistical analysis, data querying, and issues of data derivation. Data produced by large-throughput processes such as those involving microarrays or proteomics instruments require statistical mining and different forms of signal and image processing algorithms (mainly to determine clusters of different expressions). One complex aspect of data management and processing for evolutionary biology is the ability to continually curate data as new discoveries are made and to update databases based on discoveries. In evolutionary biology, unlike the other fields considered in this report, there is still no consensus on many of the mathematical models to be used, so that researchers often explore data in different ways, based on different assumptions, observations, and, importantly, the questions they are asking. Moreover, at its core, evolutionary science is founded on biological comparisons, and that means solutions to many computational problems, such as tree analyses among species or among individual organisms within and among populations, will scale in nonpolynomial time as samples increase, compounding the task of data analysis.

A two-step data-management approach seems likely in the near term for evolutionary biology, whereby individual investigators (or small teams) analyze comparatively small amounts of data, with their results and data being federated into massive data warehouses to be shared and analyzed by a larger community. The federated data will require HECC resources. (Some individual investigators will also work with massive amounts of data and HECC-enabled research, as in astrophysics and atmospheric science, but that will probably not be a common investigative paradigm for some time.) A serious concern is that the lack of tools and infrastructure will keep us from managing and capitalizing on these massive data warehouses, thus limiting progress in evolutionary biology. In particular, real-time interaction with HECC systems is required in order for biologists to discover knowledge from massive amounts of data and then use the results to guide improvements to the models. Special-purpose configurable hardware such as field-programmable gate arrays (FPGAs) might be very useful for enhancing the performance of these tasks. Finally, because evolutionary biologists are not concentrated in a few research centers, Web-based tools that allow the use of HECC systems for data analysis would be very useful.

CROSSCUTTING CHALLENGES RELATED TO EDUCATION AND TRAINING

Computational modeling and simulation are increasingly important activities in modern science and engineering, as illustrated by our four fields. As highlighted in the President's Information Technology Advisory Committee report, "computational science—the use of advanced computing capabilities to understand and solve complex problems—has become critical to scientific leadership, economic competitiveness, and national security."[1] As such, the readiness of the workforce to use HECC is seen as a limiting factor, and further investment in education and training programs in computational science will be needed in all core science disciplines where HECC currently plays, or will increasingly play, a significant role in meeting the major challenges identified in this report.

Areas that warrant education and training can be identified by considering some of the early successes of the four fields examined in this report. The atmospheric science community has evolved in the direction of well-supported community codes largely because agencies and institutions recognized that computing for both research and operations was increasingly converging toward shared goals and strategies. One consequence is that the atmospheric sciences field has offered proportionally greater opportunities for training workshops, internships, and fellowships in the computational sciences than the other three fields. The chemical separations and astrophysics communities are largely self-sufficient

---

[1]Computational Science: Ensuring America's Competitiveness, June 2005, p. iii. Available online at http://www.nitrd.gov/pitac/reports/20050609_computational/computational.pdf.

in computational science, as illustrated by the development and maintenance of robust academic codes. This stems in part from the broad training in mathematics and physical sciences received by scientists in those fields and the strong reward system (career positions in academia, government, and industry) that allows theory and computational science to thrive within the fields. Evolutionary biology has successfully collaborated with statisticians, physicists, and computer scientists, but only since the 1990s, to address the inevitable computational issues surrounding an explosion of genomic-scale data. This has pushed evolutionary biology into quantitative directions as never before, through new training grants and genuinely interdisciplinary programs in computational biology.

These early successes also show that the four fields examined here are developing at different paces toward the solution of their HECC-dependent major challenges. Evolutionary biology is increasingly moving to quantitative models that organize and query a flood of information on species, in varying formats such as flat files of DNA sequences to complex visual imagery data. To approach these problems in the future, students will need stronger foundations in discrete mathematics and statistics, familiarity with tree-search and combinatorial optimization algorithms, and better understanding of the data mining, warehousing, and visualization techniques that have developed in computer science and information technology. Evolutionary biology is currently limited by the artificial separation of the biological sciences from quantitative preparation in mathematics and physics that are standard in the so-called "hard" sciences and in engineering disciplines. Thus, the key change that must take place in evolutionary biology is greater reliance on HECC for overcoming the major challenges.

Advances in chemical separations rely heavily on compute-bound algorithms of electronic structure theory solved by advanced linear algebra techniques, as well as on advanced sampling methods founded on statistical mechanics, to do large particle simulations of materials at relevant thermodynamic state points. Chemistry and chemical engineering departments at universities traditionally employ theoretical/computational scientists who focus broadly on materials science applications but less on chemical separations problems, which are more strongly centered in the industrial sector. Moreover, these academic departments have emphasized coursework on chemical fundamentals and analytic models, and they need to better integrate numerical approaches to solving chemically complex problems into their undergraduate and graduate curricula. The chemical separations industry should consider sponsoring workshops, internships, and master's degree programs in computational science to support R&D in this economically important field.

Astrophysics and the atmospheric sciences are the most HECC-ready of the four fields, evidenced in part by their consensus on many models and algorithms. In the case of the atmospheric sciences, the community has even evolved standardized community codes. However, the chemistry and physics knowledge in these fields continues to increase in complexity, and algorithms to incorporate all of that complexity are either unknown or limited by software deployment on advanced hardware architectures. Even though these disciplines have a strong tradition as consumers of HECC hardware resources, preparation in basic computational science practices (algorithms, software, hardware) is not specifically addressed in the graduate curriculum.

Astrophysics, chemical separations, evolutionary biology, and—to a lesser extent—the atmospheric sciences typify the academic model for the development of a large-scale and complex software project in which Ph.D. students integrate a new physics model or algorithm into a larger existing software infrastructure. The emphasis in the science disciplinary approach is a proof-of-principle piece of software, with less emphasis on "hardening" the software or making it extensible to larger problems or to alternative computing architectures. This is a practical outcome of two factors: (1) limitations in training in computational science and engineering and (2) the finite time of a Ph.D. track. While computer time at the large supercomputing resource centers is readily available and no great effort is needed to obtain a block of time, software developed in-house often translates poorly to evolving chip designs on the massively parallel architectures of today's national resource machines. In response, a field typically develops its own computational leaders who, purely as a service to their communities, create more robust software and reliable implementations that are tailored to their needs. Those leaders will emerge only if there are appropriate reward mechanisms of career advancement, starting with the support for education and training to *learn* how to best advance their particular areas of computational science, especially in those science fields where computational readiness is still emerging.

Two models for education and training are being used to advance the computational capabilities of our future workforce: the expansion of computational sciences within existing core disciplines and the development of a distinct undergraduate major or graduate training as a "computational technologist." The first of these models faces the problem of expanding training and educational opportunities at the interface of the science field with computational and mathematical approaches, and to do this coherently within the time frame typical of a B.S. or Ph.D. degree. It would require integrating computational science topics into existing courses in core disciplines and creating new computational science courses that address current gaps in coverage in degree programs, which in turn call for flexibility in curricula and appropriate faculty incentives. With this approach, the rate at which standardized algorithms and improved software strategies developed by numerical analysts and computer scientists filter into the particular science is likely to be slowed. If education and training exist in undergraduate and graduate programs in a given science and engineering discipline, this is the most common model because it leads to a well-defined career path for the computational scientist within the discipline.

The second model is to develop new academic programs in computational science and engineering that emphasize the concepts involved in starting from a physical problem in science and engineering and developing successful approximations for a physical, mathematical, analytic, discrete, or object model. The student would become robustly trained in linear algebra and partial differential equations, finite difference and finite element methods, particle methods (Monte Carlo and molecular dynamics), and other numerical areas of contemporary interest. A possible limitation is that standardized algorithms and software that are routinely available may need tailoring to suit the needs of the particular science field, which only a field expert can envision. This model is the less common since the reward system for the excellent work of a computational scientist is diluted across multiple scientific/engineering disciplines and because of inherent prejudices in some scientific fields that favor insight over implementation. The primary challenge is to define a career track for a computational generalist who can move smoothly into and out of science domains as the need arises for his or her expertise and have those contributions integrated into a departmental home that recognizes their value.

Astrophysics, atmospheric sciences, and chemical separations are most ready for the first model—direct integration of computational science into the core discipline curriculum—while evolutionary biology has received the attention of statisticians, physicists, and computer scientists to develop something closer to the second model. Both models require expertise in large-scale simulation, such as efficient algorithms, data mining and visualization, parallel computing, and coding and hardware implementation. Ultimately both models will benefit any science field since both are drivers for the cross-disciplinary activity that is to be encouraged for the growing interdisciplinarity of science and engineering.

7

Conclusions

SUPPORTING HIGH-END COMPUTATIONAL RESEARCH

Variability of High-End Capability Computing

The four fields of science and engineering discussed in Chapters 2-5, while differing in many ways, all face key research challenges that cannot be addressed well or at all without new computational capabilities. In astrophysics, the committee concluded that four of the six major challenges identified are critically dependent on advances in computing capabilities, and the remaining two will require HECC to exploit the massive amounts of data that will soon be collected. In atmospheric sciences, the committee identified 10 major challenges, half of which are clearly limited by today's capabilities for computing; the other half of the 10 challenges are also impeded by limitations in HECC, but they require commensurate advances in complementary directions as well. Two of the three major challenges in chemical separations are dependent on increased application of high-end computing as well as on the development of new capabilities, although experimental approaches continue to be viable. Evolutionary biology is still an emerging field in terms of computational sophistication, and progress will no doubt take place on different fronts. Of the seven major challenges in evolutionary biology identified by the committee, the first three are ripe to benefit from increased application of high-end computing as well as from the development of new capabilities, and HECC will become essential in those areas as research pushes toward more realistic models that build on the rapidly expanding universe of data.

Progress on some of the major challenges identified in Chapters 2-5 would ensue immediately if more powerful computers and/or algorithms were available, and users would see tangible advances. For example, numerical weather predictions could be run with finer resolution, and some simulations in astrophysics and climate modeling could include additional processes or details of importance. For most of the major challenges, progress would not be so immediate, but that does not imply that investing now in appropriate HECC infrastructure is less important. For some of the major challenges, HECC investments would be particularly timely because investments have already been made in new data sources that will soon stimulate or require advances in computing.

Evaluating the potential impact of HECC in this way necessitates a definition of high-end capability computing that differs from the usual platform-centric one. From the perspective of the scientists and engineers who are working to push the frontiers of knowledge in their fields, the computational capabilities they seek are those that enable new scientific insights. Those computational capabilities serve as a lever to pry new insight from a mass of data or a complicated mathematical model. The capabilities are not simply processing "muscle," so they are not necessarily measured in terms of floating point operations per second or number of processors. Rather, from a researcher's perspective, the high-end computational capabilities include whatever mix of hardware, models, algorithms, software, intellectual capacity, and computational infrastructure must be combined to enable the desired computations. High-end computing platforms are certainly part of that mix, but ambitious and progressive computational science and engineering is a systems process that depends on many factors. Thus the committee reached the following conclusion:

Conclusion 1. High-end capability computing (HECC) is advanced computing that pushes the bounds of what is computationally feasible. Because it requires a system of interdependent components and because the mix of critical-path elements varies from field to field, HECC should not be defined simply by the type of computing platform being used. It is nonroutine in the sense that it requires innovation and poses technology risks in addition to the risks normally associated with any research endeavor.

Infrastructure Needs of HECC

At the very least, HECC infrastructure consists of hardware, operating software, and applications software. There is also a need for data management tools, graphical interface tools, data analysis tools, and algorithms research and development. Some critical problem areas may need targeted research into mathematical models, and others might require training or incentives to speed a targeted community's climb up the learning curve. All parts of the HECC ecosystem must be healthy in order for HECC-enabled research to thrive. Some high-opportunity fields will not fully exploit HECC unless other changes fall into place. An example is chemical separations: While there is already a strong community of computational chemists skilled at HECC, they are not generally working on problems of industrial importance, for a variety of valid reasons. If it is determined that scientific progress is impaired because of underexploitation of HECC, then the incentives that drive computational chemistry researchers should be changed.

Fields will generally take advantage of the increased availability of HECC in proportion to how much of the necessary infrastructure has been created—that is, whether the field is ready for HECC. All available evidence suggests that the advancement of science and its applications to society increasingly depends on computing. For all fields to contribute, they must receive support for whatever they need to ready them to capitalize on HECC. The committee members, even though very diverse in disciplinary and computing expertise, readily reached the following conclusion:

Conclusion 2. Advanced computational science and engineering is a complex enterprise that requires models, algorithms, software, hardware, facilities, education and training, and a community of researchers attuned to its special needs. Computational capabilities in different fields of science and engineering are limited in different ways, and each field will require a different set of investments before it can use HECC to overcome the field's major challenges.

Drivers for Investment Decisions

Once a decision is made to invest in computational resources, that investment must address all the elements of infrastructure that are needed by fields likely to use the resource. HECC infrastructure should be interpreted to mean whatever set of investments is needed to enable the desired progress. Every element that contributes to computational capability should be of concern to the providers of computing infrastructure. The needs vary from field to field, and optimal progress in science and engineering is not likely unless HECC infrastructure suits the existing capabilities of the fields it is meant to serve.

What are the preconditions for a field to profit from HECC? At a minimum they would include the following:

- The field must have established mathematical models for important research questions.
- Algorithms must exist or be under development for computing solutions to those models.
- The field should have sufficient theory and experimental data on which to base the models, if not to completely validate them.
- Some simplified computations should already have been performed, so that the value and limits of computational approaches are becoming clear.
- The relevant community must see value in computational approaches.
- The research community must have or be able to tap into appropriate skills in computation, including the ability to optimize models and algorithms for HECC architectures.
- Appropriate hardware and computational software must be available and convenient.
- Some community software should be available or, alternatively, researchers using HECC should be well aware of how other computational scientists in their field are approaching similar problems.
- Supporting software (for example, data management tools, automated grid generators, interface software) must be available, suitable, and understood by the researchers.
- There must be enough of a peer community to adequately review and discuss computational results and to afford students working on computational research a viable career track.

Conclusion 3. *Decisions about when, and how, to invest in HECC should be driven by the potential for those investments to enable or accelerate progress on the major challenges in one or more fields of science and engineering.*

THE NEED FOR CONTINUING INVESTMENT IN HECC

Task (d) of the Statement of Task for this study calls for the committee to reflect on the opportunity costs of simply waiting for computing capabilities to improve in response to some of the same competitive incentives that have been driving information technology for several decades now. Some of the major challenges identified in this report, particularly those that would enable better understanding of climate change, are innately urgent, so that delays in addressing them would affect more than just scientific progress. More generally, the committee believes that the major challenges identified in this report cannot necessarily be decoupled from one another and supported selectively, which led the committee to the following conclusion:

Conclusion 4. *Because the major challenges of any field of science or engineering are by definition critical to the progress of the field, underinvestment in any of them will hold back the field.*

Optimum progress will be achieved if all modalities of research—theoretical, experimental, and computational—are supported in a balanced way. In many cases, HECC capabilities must continue to advance in order to maximize the value of data already collected or experimental investments already made or committed. For instance, remote-sensing projects under way in astrophysics and atmospheric science will produce quantities of data that cannot be utilized by those fields without commensurate progress in analytical capabilities. In evolutionary biology, so much genomic information is being generated that the paucity of tools for handling the information in a comparative framework is jeopardizing its value; in short, the inflow of observations is exceeding our ability to process them. We want to avoid what is known as the "write-once, read-never" phenomenon. The value of massive amounts of data cannot be properly leveraged without commensurate investments in the HECC infrastructure needed for their analysis.

Most of the opportunities opened up by HECC, as described in this report, require long-term R&D. For those that will require the next-generation of computer architectures, many algorithmic and software developments must come together before the opportunities can be realized. Computers, models, applications, and knowledge all evolve in a coordinated way. If researchers simply watch the advance of computing technology and wait until it possesses the needed capabilities, they will not be prepared to use the new capabilities effectively and efficiently. Past experience—adapting first to vector processors and then to parallel processors—suggests that this will entail years of effort. For those opportunities that require other sorts of groundwork, such as model development, there is also a long path toward capitalizing on the opportunities. To minimize the risk of technological surprise, multiple directions must be tried out—and tried out now. The issue raised by Task (d) is not just whether to invest in a new platform, because so many other steps must in any case come together before a new capability is available. The real issue is how to stage those preparations so as to build new capability. The risk of not investing now is that those steps will not be ready when needed, and that technological risks will be too high.

The committee notes that Task (d) seems to assume that HECC is focused on the computing platform, with jumps in capability being tied to advances in hardware. That is somewhat at odds with the more holistic view expressed in committee Conclusions 1 and 2. In the committee's view, different fields must take different steps in order to capitalize on HECC, and many of those steps do not depend on pushing the state of the art in processing speed. This is not to say that there is never a need for the federal government to invest in computing hardware that is more ambitious than that targeted by the commercial world. It is, rather, to drive home the point that hardware investments are not the only way for the federal government to advance HECC.

The premise behind Task (d) is that the information technology industry has for several decades advanced processing capabilities exponentially. To continue at that rate, which is generally referred to as following Moore's law, computer manufacturers are now moving toward new hardware architectures, as yet undefined. No matter what form these architectures take, this change is likely to be very disruptive, akin to the shift from von Neumann to parallel architectures in the 1980s. That shift required a great deal of effort in algorithm and software research before users could effectively use many processors in parallel. The emergence of new architectures will probably necessitate a comparable amount of research to develop robust and efficient basic algorithms that make good use of the new architecture. An exponential increase in computing speed is not guaranteed, because we cannot assume that existing codes will port well to the new platforms.

There are still no productive and easy-to-use programming methodologies or low-level blocks of code that can take full advantage of multicore processors. Multicore parallelism is unfamiliar to many commercial software developers, and it also requires different sorts of parallel algorithm development. For instance, MPI and Open-MP, which are common software packages used throughout astrophysics,

CONCLUSIONS   *125*

will not work well on multicore architectures. Because the hardware architecture for the next generation of HECC machines is not yet defined, efficient software libraries have yet to be developed. Moreover, computer-science education has focused on teaching sequential algorithms, while automated methods, such as those in compilers, cannot deduce algorithmic concurrency from most sequential codes. Many algorithms will have to be rethought and much software rewritten.

As a result of these impending developments, the committee reached the following conclusion:

Conclusion 5. *The emergence of new hardware architectures precludes the option of just waiting for faster machines and then porting existing codes to them. The algorithms and software in those codes must be reworked.*

CLASSES OF NUMERICAL AND ALGORITHMIC CHALLENGES

Chapter 6 discusses the classes of numerical and algorithmic challenges that the committee discerned from the four areas of science and engineering covered by this study. The committee reached the following conclusion about the classes of numerical and algorithmic characteristics that will be needed:

Conclusion 6. *All four fields will need new, well-posed mathematical models to enable HECC approaches to their major challenges. Astrophysics and the atmospheric sciences share two needs: one for new ways to handle stiff differential equations and one for continuing advances in multi-resolution and adaptive discretization methods. Astrophysics and chemical separations also share two needs: one for accurate and efficient methods for evaluating long-range potentials that scale to large numbers of particles and processors and one for stiff integration methods for large systems of particles.*

The HECC-dependent challenges in all four of the fields studied will rely on software of much greater complexity than that currently in use, and it must be optimized for new (as-yet-undefined) computer architectures. Both factors point to the need for large, sustained efforts in software.

In addition, it is very clear that data management, analysis, and mining are increasingly critical and crosscutting algorithmic challenges. While data-intensive computing is not always thought of as within the province of HECC, it is clear that high-end computational science and engineering very often stress data-management capabilities. Very powerful computer simulations can generate hundreds of gigabytes of data, which are very difficult to manage and visualize. Other types of research are faced with similarly enormous sets of input data, such as those from satellites and telescopes, and in those situations HECC capabilities are required to digest data and create insight.

HUMAN RESOURCES

Conclusion 5 implies that the committee foresees an increasing need for computational scientists and engineers who can work with mathematicians and computer scientists to develop next-generation HECC software. Chapters 4, 5, and 6 explicitly mention the need for more widespread education about scientific computing. A typical earth scientist, for instance, is not prepared to transition code from an IBM supercomputer to a Cray. That is a task that calls for specialists with a knowledge of software engineering, applied mathematics, and also some domain knowledge. In the atmospheric sciences, it might be feasible to plug in new physics models and not risk disturbing the underlying performance of the code, but in other fields, like computational chemistry, one needs to dig deeply into the code and model in order to work on its optimization.

Based on these observations, the committee reached the following conclusion:

Conclusion 7. *To capitalize on HECC's promise for overcoming the major challenges in many fields, there is a need for students in those fields, graduate and undergraduate, who can contribute to HECC-enabled research and for more researchers with strong skills in HECC.*

The career path for people who invest time in developing high-end computing capabilities, which by themselves might not constitute publishable research, is problematic, especially in academia. What is needed is a career path that encompasses both a service role (HECC consulting within their field and to computer scientists) plus opportunities to conduct their own research.

LESSONS LEARNED FOR FIELDS THAT MIGHT PERFORM SIMILAR STUDIES

The findings and conclusions in this chapter might apply to other fields of science and engineering, but the committee did not explore that question. The study that led to this report was, in part, an experiment to determine whether particular fields of science and engineering could follow a methodology known as "gaps analysis" to determine the potential impact of—and hence their implied need for—advanced computing. In the committee's view, the experiment was a success. Even though the four fields selected for this study are very disparate, the committee was able to develop credible snapshots of the major challenges from each of those fields and then determine which of them are critically dependent on HECC. Any other field that wishes to perform a similar self-assessment should take the following lessons to heart:

- It is necessary to build on existing statements about a field's current frontiers or major challenges. Developing a consensus picture of the frontier, and of the major challenges that define promising directions for extending that frontier, is a major task by itself.
- It is important to determine which major challenges for the field are critically dependent on HECC. It is easy to spot opportunities for applying HECC to advantage, but that is not the same as identifying the major challenges where progress will be limited if appropriate HECC cannot be brought to bear.
- Compelling justification for a particular proposed HECC investment would require a level of analysis not included in this report. For each of the major challenges targeted by the investment, it would be appropriate to identify the various research directions that are germane to progress and their associated infrastructure requirements. From them, one could develop an investment strategy that maximizes the potential for scientific progress.

All the infrastructure components needed to apply HECC to the challenges that depend on it must be identified, and the community must develop a clear understanding of the resources needed to build a complete infrastructure. Merely giving a field access to supercomputers is no guarantee that the field's scientific progress will be enabled or accelerated.

Appendixes

Appendix A

Biographical Sketches of Committee Members

John W. Lyons, *Chair*, National Defense University, has directed two major federal science and engineering laboratories. He is a physical chemist with degrees from Harvard College and Washington University in St. Louis, Missouri. He began his career in research and development positions with the Monsanto Company for 18 years. In 1973 he joined the Commerce Department's National Bureau of Standards (NBS) at Gaithersburg, Maryland. At NBS Dr. Lyons was the first director of the Center for Fire Research and then in 1978 the first director of the National Engineering Laboratory, a unit that came to include about half of the NBS programs. In 1990, Dr. Lyons was appointed by President George H.W. Bush to be the ninth director of NBS, by that time renamed the National Institute of Standards and Technology (NIST). In September 1993, he was appointed the first permanent director of the U.S. Army Research Laboratory (ARL). At ARL, Dr. Lyons managed a broad array of science and technology programs. Dr. Lyons has published four books and over 60 papers, and he holds a dozen patents. He was elected to the National Academy of Engineering in 1985. He is a fellow of the American Association for the Advancement of Science and of the Washington Academy of Sciences and a member of the American Chemical Society and of Sigma Xi.

David Arnett is professor of astrophysics at the Steward Observatory of the University of Arizona. He is a theoretical astrophysicist who first demonstrated how explosive nucleosynthesis in supernovae produces the elements from carbon through iron and nickel. He constructed quantitative theoretical models of evolving massive stars and showed that the ejecta produce a good fit to the abundance of heavy elements in the galaxy. His research interests include nuclear astrophysics, formation of neutron stars and black holes, high-performance computers, theoretical physics, hydrodynamics, thermonuclear burning, stellar evolution, computer graphics, and computer modeling. Dr. Arnett is a member of the National Academy of Sciences.

Alok N. Choudhary is chair of the Electrical and Computer Engineering Department at Northwestern University. He also holds an adjunct appointment with the Kellogg School of Management in marketing and technology innovation. In 2000, he cofounded Accelchip, Inc., a developer of electronic design automation tools and services. Dr. Choudhary is the founder and director of the Center for Ultra-scale Computing and Information Security (CUCIS). His research interests are in high-performance computing and communication systems, power aware systems, computer architecture, and high-performance I/O systems and software, and their applications in many domains including information processing and scientific computing. His interests include the design and evaluation of architectures and software systems, high-performance servers, high-performance databases, and input-output and software protection/security.

Phillip Colella is a senior mathematician and group leader of the Applied Numerical Algorithms Group (ANAG) at Lawrence Berkeley National Laboratory. The mission of ANAG is the development of advanced numerical algorithms and software for partial differential equations integrated with the application of the software to problems of independent scientific and engineering interest. The primary focus of its work is the development of high-resolution and adaptive finite difference methods for partial differential equations in complex geometries, with applications to internal combustion engines and other industrial problems. In 1998, Dr. Colella was awarded the Sidney Fernbach Award from the IEEE Computer Society for his outstanding work in numerical algorithm development and parallel code design and implementation. The award is given annually to computational scientists who have achieved breakthroughs in high-performance computing. Dr. Colella's award cites his "fundamental contributions in the development of software methodologies used to solve numerical partial differential equations, and their application to substantially expand our understanding of shock physics and other fluid dynamics problems." Dr. Colella is a member of the National Academy of Sciences.

Joel L. Cracraft is Lamont Curator of Birds and curator in charge of the Division of Vertebrate Zoology and Ornithology at the American Museum of Natural History. He is also an adjunct professor in the Department of Earth and Environmental Sciences at Columbia University and in the Department of Biology at the City University of New York. His research interests include diversification in birds, diversification and evolution, and systematic and biogeographic theory and methods.

John A. Dutton is professor emeritus in the College of Earth and Mineral Sciences at the Pennsylvania State University and a principal in Storm Exchange, Inc. His research interests involve dynamic meteorology, including dynamical systems, spectral models, predictability, climate theory, and global change. Dr. Dutton's interests span a number of topics in nonlinear atmospheric dynamics, with a current focus on the properties of attractors of hydrodynamical systems, on problems in predictability, and on global properties of atmospheric flow.

Scott V. Edwards is a professor in the Department of Organismic and Evolutionary Biology at Harvard University. He studies the evolutionary biology of birds and their relatives, combining field, museum, and genomics approaches to understand the basis of avian diversity, evolution, and behavior. His research involves population genetics; geographic variation and genome evolution; and systematics. He is a past president of the Society of Systematic Biologists.

David J. Erickson III is a senior research staff member and director of climate and carbon research in the Center for Computational Sciences at Oak Ridge National Laboratory. He is also an adjunct professor in the Division of Earth and Ocean Sciences, Nicholas School of the Environment and Earth Sciences, Duke University. Dr. Erickson's research interests include global climate modeling, numerical modeling of atmospheric chemistry, and modeling the global air-sea exchange of energy, momentum, trace gases, and particles.

Teresa L. Head-Gordon is an associate professor in the Department of Bioengineering at the University of California at Berkeley. Her research program encompasses the development of general computational and experimental methodologies applied to chemistry and biology in areas such as protein aggregation disease, biomaterials assembly, and glassy dynamics of nanomaterials. She is the recipient of an IBM SUR award (2001) and was Schlumberger Visiting Professor at Cambridge University in 2005-2006. Dr. Head-Gordon serves as editorial advisory board member for the *Journal of Computational Chemistry* and editorial board member for the SIAM book series on computational science and engineering (2004-present).

Lars E. Hernquist is professor and chair of the Harvard-Smithsonian Center for Astrophysics. His research interests include theoretical studies of dynamical processes in cosmology and galaxy formation and galaxy evolution, numerical simulations of stellar dynamical and hydrodynamical systems, and investigations of the physics of compact objects, particularly neutron stars and the interplay between thermal and magnetic processes in strongly magnetized neutron stars. Prof. Hernquist is a member of the National Academy of Sciences.

George E. Keller II is retired senior corporate research fellow at the Union Carbide Corporation and now vice chairman of the Mid-Atlantic Technology, Research and Innovation Center (MATRIC), an independent, nonprofit, 501(c)(3) corporation headquartered in West Virginia. He is noted for invention and insightful analysis of novel separation processes. His expertise is in chemical and petroleum separation technologies, including distillation, membranes, adsorption, and extraction, and he is coauthor of the book *Separation Process Technology*. Dr. Keller is a member of the National Academy of Engineering.

Nipam H. Patel is Howard Hughes Investigator and professor of genetics and development and of integrative biology in the Department of Molecular and Cell Biology of the University of California at Berkeley. His research program centers on the study of the evolution of development mechanisms with a focus on the genes that regulate segmentation and regionalization of the body plan. He is particularly interested in understanding how certain steps in patterns formation that require protein diffusion in *Drosophila* are accomplished in those insects and crustaceans in which cellularization of the growing embryos would seem to preclude formation of gradients by diffusion. His group also investigates the role of homeotic genes in generating body plan diversification in crustaceans. He is also investigating the function of the *Drosophila* segmentation genes during neuronal development and how they may have contributed to the evolution of neural complexity.

Mary E. Rezac is professor and head of the Department of Chemical Engineering at Kansas State University. Her fields of research include mass transport, polymer science, membrane separation processes, hybrid system (reactor-separator) designs, and applications to biological systems, environmental control, and novel materials.

Ronald B. Smith is professor of geology and geophysics and of mechanical engineering at Yale University and director of the Yale Center for Earth Observation. He leads Yale's program in mesoscale meteorology and regional climate, which includes atmospheric dynamics, observations of the atmosphere using aircraft and satellite, hydrometeorology using stable isotopes of water and theories of evaporation and rain, and satellite remote sensing of landscape changes and climate sensitivity.

James M. Stone is a professor in the Department of Astrophysical Sciences at Princeton University, with a joint appointment in the Program in Applied and Computational Mathematics. His research group studies gas dynamics in a wide variety of astrophysical systems, from protostars to clusters of galaxies. As part of this effort, the group develops, tests, and applies numerical algorithms for astrophysical gas dynamics on high-performance computers.

John C. Wooley is associate vice chancellor for research at the University of California at San Diego (UCSD), where he is also an adjunct professor in the Department of Chemistry and Biochemistry and in the Department of Pharmacology at the School of Medicine. He is also a research associate professor of biophysics at John Hopkins Medical School, a member of the National Institutes of Health's (NIH's) Resource for Macromolecular Modeling and Bioinformatics, director of NSF's Biological Sciences Advisory Committee, and a member of the advisory committee for the National Biomedical Computation Resource. Prior to moving to UCSD, Dr. Wooley spent time in government service at NSF, NIH, and DOE. His research focuses on structure-function relationships in protein-nuclei acid complexes and the architecture of chromatin and ribonucleoproteins. He collaborated on the first stages of the Human Genome Project and established the first federal programs in bioinformatics and computational biology. As associate vice chancellor, he is also taking the lead in a variety of biotechnology and computational biology projects, including centers for structural genomics, bioinformatics, cell signaling, biomimetic materials, and computation science/distributed computing. In general, he is creating and facilitating new interdisciplinary research and education efforts that cross traditional interdisciplinary boundaries and new scientific teams from the university and partner institutions.

Appendix B

Agendas of Committee Meetings

MEETING 1

WASHINGTON, D.C.
SEPTEMBER 29-30, 2006

Friday, September 29

Closed Session

8:00-9:30 a.m.

Open Session

9:30	Purpose of the study and how it will be used John H. Marburger III, director of the White House Office of Science and Technology Policy (OSTP)
10:00	Need for the study, as seen by the White House Office of Management and Budget (OMB) Joel Parriott, OMB
10:30	Break
10:45	Overview of the federal Networking and Information Technology Research and Development (NITRD) program and its National Coordination Office (NCO) Simon Szykman, NCO/NITRD director

11:15	Discussion of study tasks and scope George Strawn, NSF chief information officer and chair of NITRD steering group for the study (presenter) Sally Howe, NCO/NITRD associate director
12:15 p.m.	Working lunch
1:00	Initial thoughts from the computational science and engineering subgroup of the committee (Choudhary, Colella, Head-Gordon, Wooley) on the context, scope, and goals of the study. Discussion leader: Phil Colella, Lawrence Berkeley National Laboratory
1:45	Plenary discussion of the charge, definitions, scope, plan of work, audience(s), and desired outcomes of the study and desired outcomes of the study
2:45	Break
3:00	Overview of the important scientific and technological problems in astrophysics Christopher McKee, University of California at Berkeley (by speakerphone)
4:00	NSF perspectives on the study Arden Bement, NSF director Note: NSF's perspective is included as an example of an agency that relies on high-end computing. This does not imply that the study will focus preferentially on NSF interests.
4:30	Open discussion of the role of high-end computing in science and engineering
5:15	Reception (guests invited)
6:15	Adjourn

Saturday, September 30

Closed Session

8:00 a.m. to
3:00 p.m.

APPENDIX B   *135*

MEETING 2

WASHINGTON, D.C.
DECEMBER 14-15, 2006

Thursday, December 14

Open Session

8:00 a.m.   Working breakfast (entire committee and all guests)

8:30   Discuss the study charge and the goals of the parallel workshops (Lyons)

8:55   Break into four concurrent workshops (check room assignments). The workshops will follow the agendas below from 9:00 a.m. until 4:30 p.m., when a plenary sesion will take place.

Four Concurrent Workshops:
Astrophysics, the Atmospheric Sciences,
Chemical Separations, and Evolutionary Biology

Define the Science/Engineering Requirements

9:00   Presentation(s) or discussion to identify the major research questions for the field (independent of the role computing may or may not play)

10:15   Break

10:30   Reach consensus on the list of major research questions that should be considered by the study and identify any documents (e.g., consensus reports) that support inclusion of items on the list.

11:00   For which of these major research questions would high-end modeling, simulation, and analysis play an irreplaceable role? List this subset of research challenges, which will be referred to as "computation-dependent challenges." Describe what could be accomplished with high-end computing and what kind(s) of infrastructure is (are) needed. Could other techniques (e.g., experiments, observation, more traditional theory) serve as a substitute? If this high-end capability is not proactively developed, what would be lost?

12:15 p.m.   Working lunch in breakout rooms

Examine Current Capabilities for High-End Computational Approaches to the Computation-Dependent Challenges

1:00   Evaluate the current status of high-end modeling, simulation, and analysis for the computation-dependent challenges. Consider the state of the mathematical models, algorithms, and software. Comment on the past track record of computational approaches to these challenges. For each of these computation-dependent challenges, consider the following questions:

- Are there strong experimental programs to inform a computational approach, so that hypotheses are testable and investigators can agree on the knowledge gaps?
- Are there large experimental data repositories?
- Is there funding for research programs that combine theory and experiment, conferences and publications that bring them together, and other evidence of a healthy interplay?
- Does knowledge about the particular challenge extend well beyond classification/organization and proceed toward well-defined computational models?
- Is the relevant community poised to investigate the particular challenge through high-end computing? For example, is the infrastructure (hardware, community software, support) available and used?
- Is there a demand for computational scientists in the field from which this particular challenge arises? For example, would computation-intensive graduate students be employable?

2:15   Identify possible barriers to successfully addressing these computation-dependent challenges, considering the adequacy of models, algorithms, software, and computer resources. Cluster these barriers into those that are of near-term concern and those that are longer-term challenges.

2:45   Break

Examine New Capabilities for High-End Computational Approaches to the Computation-Dependent Challenges

3:00   What new technologies (broadly defined) could be applied to overcome the barriers identified at the end of the last session?

3:45   What would be the effort required, the risk, and the time horizon for successful application of these new technologies?

Astrophysics Workshop Participants

Committee members
Dave Arnett, University of Arizona
Alok Choudhary, Northwestern University
Jim Stone, Princeton University

Guests
Tom Abel, Stanford University
Eve Ostriker, University of Maryland (afternoon only)
Ed Seidel, Louisiana State University
Nigel Sharp, NSF
Alex Szalay, Johns Hopkins University

Staff
Jim McGee, National Research Council

Atmospheric Sciences Workshop Participants

Committee members
Phil Colella, Lawrence Berkeley National Laboratory
John Dutton, Pennsylvania State University
Ron Smith, Yale University

Guests
Antonio J. Busalacchi, University of Maryland
Brian Gross, NOAA Geophysical Fluid Dynamics Laboratory
James Hack, NCAR
Chaowei (Phil) Yang, NASA Applied Sciences Program

Staff
Neal Glassman, National Research Council

Chemical Separations Workshop Participants

Committee members
George Keller, Mid-Atlantic Technology, Research and Innovation Center (MATRIC)
John Lyons, Army Research Laboratory (retired)[1]
Mary Rezac, Kansas State University

Guests
Joan Brennecke, University of Notre Dame
Anne Chaka, NIST
Thom Dunning, University of Illinois at Urbana-Champaign
Jeff Siirola, Eastman Chemical

Staff
Scott Weidman, National Research Council

---

[1] Dr. Lyons will spend time in each of the workshops.

Evolutionary Biology Workshop Participants

Committee members
Joel Cracraft, American Museum of Natural History
Scott Edwards, Harvard University
Nipam Patel, University of California at Berkeley
John Wooley, University of California at San Diego

Guests
Daniel Drell, Department of Energy
Sean Eddy, Howard Hughes Medical Institute
Sergey Gavrilets, University of Tennessee, Knoxville
Matthew Kane, NSF
Joel Kingsolver, University of North Carolina
Daniel Rokhsar, Lawrence Berkeley National Laboratory
Rick Stevens, Argonne National Laboratory
Manfred Zorn, NSF

Staff
Ann Reid, National Research Council

Plenary Session: All Committee Members and Guests Reconvene

4:30	Brief reports from each workshop. What has been learned? How will the study use these results?
5:10	Adjourn for day

Closed Session

7:00	Committee working dinner

Friday, December 15

Open Session

8:00 a.m.	Working breakfast
8:30	Report from the astrophysics workshop
9:15	Report from the atmospheric sciences workshop
10:00	Break
10:15	Report from the chemical separations workshop
11:00	Report from the evolutionary biology workshop

APPENDIX B

Closed Session

11:45 a.m. to
3:00 p.m.

MEETING 3

IRVINE, CALIFORNIA
MARCH 8-9, 2007

Thursday, March 8

Closed Session

5:30 p.m.   Committee working dinner
to
8:30 p.m.

Friday, March 9

Open Session

8:00 a.m.   Working breakfast

8:30   Discussion of each of the disciplinary chapters

10:30   Break

Closed Session

10:45 a.m.   Discussion of drafts
to
5:30 p.m.

MEETING 4

WASHINGTON, D.C.
MAY 7-8, 2007

Closed Session

Appendix C

Glossary

Clade	Group of species
Exascale	Computing at quintillions of flops per second
Giant Molecular Cloud	A large cloud of molecular gas and small solid particles from which new generations of stars form
Heuristic	Algorithm based on one or more rules of thumb rather than on a theoretical model
Magnetohydrodynamics	Study of the dynamics of electrically conducting fluids
Morphology	External character of an organism
Multicore	Pertaining to a computer chip with multiple processing units on it, or a computer that includes such chips
Ontology	Set of defined concepts that characterize a domain
Orographic	Caused by airflow over hills or mountains
Petaflop	Quadrillion floating point operations
Petascale	Computing at quadrillions of flops per second
Phenotype	Observable characteristics of an individual

Phylogenetics	Relationships of groups of organisms as reflected by their evolutionary history
Stiff differential equations	Coupled differential equations that include some terms that lead to rapid variations; they must be handled with special solvers to avoid instabilities
Symplectic integrators	Discretizations that have a Hamiltonian structure analogous to that of the dynamics being approximated
Teraflop	Trillion floating point operations
Terascale	Computing at trillions of flops per second